Astronomers' Observing Guide

Other titles in this series

Double and Multiple Stars and How to Observe Them
James Mullaney

Related titles

Field Guide to the Deep Sky Objects
Mike Inglis

Deep Sky Observing
Steven R. Coe

The Deep-Sky Observer's Year
Grant Privett and Paul Parsons

The Practical Astronomer's Deep-Sky Companion
Jess K. Gilmour

Observing the Caldwell Objects
David Ratledge

Choosing and Using a Schmitt-Cassegrain Telescope
Rod Mollise

Peter Grego

The Moon
and How to
Observe It

With 134 Figures

 Springer

Series Editor: Dr. Mike Inglis, FRAS

British Library Cataloguing in Publication Data
Grego, Peter
    The moon and how to observe it:an advanced handbook for
    students of the moon in the 21st century.-(Astronomers'
    observing guides)
    1. Moon—Observers' manuals
    I. Title
    523.3
ISBN 1852337486

Library of Congress Cataloging-in-Publication Data
Grego, Peter.
    The moon and how to observe it: an advanced handbook for students of the moon in the
    21st century / Peter Grego.
        p. cm. —(Astronomers' observing guides, ISSN 1611-7360)
    Includes bibliographical references and index.
    ISBN 1-85233-748-6 (acid-free paper)
     1. Moon—Observers' manuals. 2. Moon—Observations. I. Title. II. Series.
    QB581.G74 2005
    523.3—dc22
                                                                                    2004052524

Astronomers' Observing Guides Series ISSN 1611-7360
ISBN 1-85233-748-6
Springer Science+Business Media
springeronline.com

Typeset by EXPO Holdings Sdn Bhd
Printed in Singapore
58/3830-543210   Printed on acid-free paper   SPIN 10930052

To my daughter Jacy and my wife Tina

Acknowledgments

Thanks to Mike Inglis for having first asked me to write this book, and for his help and advice as the project got underway. All the staff at Springer in the United Kingdom and United States have worked hard to produce this book, and I am deeply grateful to them. My special thanks go to John Watson and Louise Farkas.

All of my lunar observing acquaintances have been a tremendous inspiration to me. Thanks to Mike Brown, Doug Daniels, Chris Dignan, Colin Ebdon, Mike Goodall, Brian Jeffrey, Nigel Longshaw, Dusko Novakovic and Grahame Wheatley for their kind permission to use their hard-won lunar images and observational drawings.

Peter Grego
Birmingham, United Kingdom

Contents

Part 1 About the Moon

1 **The Moon's Origin** 3
   Fission 4
   Co-accretion 5
   Capture 6
   Collision 8
   The Moon's Development 9
   Inside the Moon 12
   Types of Lunar Rock 15
   Shaping the Moon's Surface 18
   Changes on the Moon 35

2 **The Measure of the Moon** 41
   The Moon's Orbit 42
   Gravity and Tides 44
   Secular and Periodic Perturbations 45
   Libration 46
   Lighting Effects 50
   Retardations 54
   Atmospheric Phenomena 54
   Eclipses 54
   Lunar Occultations 57
   The Search for Other Natural Satellites 58

3 **Worlds in Comparison** 60
   Mercury 60
   Venus 62
   Earth 63
   Mars 64
   The Satellites of Mars 65
   Asteroids and Comets 66
   Jupiter's Satellites 67
   Saturn's Satellites 69
   Uranus's Satellites 69
   Neptune's Satellites 70

Part 2   Observing the Moon

4   **Observing and Recording the Moon** 75
  Drawing the Moon 76
  Observational Information 81
  Imaging the Moon 85

5   **Viewing the Moon with the Unaided Eye** 98
  Features Visible on the Moon 98
  Earthshine 100
  Perception Is Not Always Reality 100
  Lunar Cross-Staff 101
  Tracking the Moon 101
  Color Perception in Moonlight 102
  Atmospheric Effects 102

6   **Lunar Showcase: A Binocular Tour of the Moon's Trophy Room** . . 104
  Maria: Vast Seas of Solified Lava 105
  Tour of Mare Imbrium: A Small Telescope Trek 108
  Wrinkle Ridges 111
  Domes 111
  Impact Craters 113
  Ray Studies 115
  Faults 116
  Observing Lunar Eclipses 119

7   **A Survey of the Moon's Near Side** 122
  Some Lunar Terminology 124
  Northeastern Quadrant 125
  Northwestern Quadrant 149
  Southwestern Quadrant 172
  Southeastern Quadrant 199

8   **Advanced Lunar Research** 225
  Transient Lunar Phenomena (TLP) 225
  Banded Lunar Craters 227
  The Measure of the Moon: Calculating Crater Depths and Mountain
    Heights 230

9   **The Lunar Observer's Equipment** 232
  An Eye for Detail 232
  Binoculars 234
  Telescopes 236
  Eyepieces 244
  Binocular Viewers 246
  Telescope Mounts 247

Glossary 250

Appendix: Resources 260

About the Author 265

Index of Lunar Features 267

Subject Index 272

Contents

Part 1

About the Moon

Chapter 1

The Moon's Origin

The Moon, planet Earth's only natural satellite, is exceedingly old, having existed as an independent body for around 4.6 billion years. One of astronomy's most enduring mysteries is how the Earth came to be partnered by such a large satellite. In comparison with the size of the planet it orbits, the Moon ranks as the Solar System's second largest satellite. At 3,476km across, the Moon measures about a quarter the diameter of the Earth (12,756km); Pluto's satellite Charon is relatively larger, measuring more than half Pluto's diameter (2,320km compared with 1,172km). By virtue of the size of their satellites, both the Earth–Moon system and the Pluto–Charon system are sometimes referred to as "double" planets. There are several physically larger satellites than the Moon – namely, Saturn's satellite Titan (5,150km) and Jupiter's satellites Io (3,630km), Ganymede (5,268km) and Callisto (4,806km) – but they themselves are utterly dwarfed by the giant planets they orbit.

It had initially been hoped that samples of lunar material studied in laboratories would provide a lucid insight into the very birth of the Moon. Although it has been relatively easy to identify their mineral composition, lunar rocks have told scientists more about the local conditions in which they were created – in most cases revealing an exceedingly complex history of melting and crystallization than showing examples of pristine lunar bedrock.

There are a number of constraints facing current theories of the Moon's origin. They must account for our satellite's relatively large size and explain why it has a near-circular orbit inclined by 5° to the Earth's orbital plane (the ecliptic). The Earth–Moon system has the largest amount of angular momentum (a product of the objects' masses and velocities) in the Solar System; what mechanisms could possibly account for this? In addition, any plausible theory must not overlook the fact that the Moon's core is proportionately far smaller than the Earth's – less than 3% its mass compared with 30% for the Earth's core. In fact, the Moon has about the same density as the Earth's mantle, but the Earth itself is 1.6 times denser than the Moon. Why the big difference in density between the Earth and Moon? Taken as a whole, the Moon possesses substantially less iron than the Earth; its rocks contain no water and they are depleted in other volatiles (substances that vaporize at relatively low temperatures). Lunar-formation theories must also account for the fact that the nonradioactive, stable isotopes of oxygen (^{16}O, ^{17}O, ^{18}O) in lunar and terrestrial rocks have identical relative abundances, implying that the Earth and Moon formed at the same distance from the Sun in the same general vicinity within the solar nebula.

Four major scientific theories have attempted to explain the Moon's origin:

1. *Fission*. The Moon is a chunk of material spun off from a rapidly revolving Earth.

2. *Co-accretion* (also known as the 'Sister Planet' theory). The Moon formed from a cloud of debris in orbit around the embryonic Earth.
3. *Capture.* The Moon originally formed as an independent planet but was captured by the Earth's gravity.
4. *Collision* (also known as the 'Big Whack' theory). The Moon was formed from material thrown out from the Earth after a planetary collision.

Fission

The fission theory states that the material of the Moon and Earth was once incorporated in a single proto-planet that accreted from material within the solar nebula around 4.6 billion years ago. After rapidly becoming a hot, internally molten globe because of the rapid accumulation of material and radioactive decay, the heavier metallic elements sank down through the outer layers of the proto-planet's mantle to form an iron-rich core. Because the inner regions of the proto-planet became more massive, there was a gradual acceleration in its axial rotation, and centrifugal forces gave rise a pronounced equatorial bulge. Instability within the proto-planet's rapidly rotating molten mass, produced mainly by the gravitational pull of the Sun, caused it to elongate. A nodule eventually broke away, and this became the embryonic Moon.

The fission theory originally developed from the fact that the Moon is gradually moving away from the Earth (at a current rate of 4cm per year). As the Moon raises tidal bulges in the Earth's oceans, friction between the ocean masses and the Earth's crust acts like a brake, gradually decelerating the Earth's rate of spin. Friction causes the tidal bulges to be displaced forward of the line joining the

Fig. 1.1. Now disfavored, the fission hypothesis of the formation of the Moon was first proposed in the 19th century. It saw the Earth and Moon as having once been joined in a hot rotating body whose rapid spin created centrifugal forces powerful enough to fling off the material of the Moon. Credit: Peter Grego

center of the Earth and Moon, and it is the gravitational attraction of the displaced tidal bulge facing the Moon that is tugging on it, imparting additional angular velocity; this in turn results in a gradual spiraling out of the Moon's orbit. Tidal forces have the net effect of slowing the Earth's rotation by a tiny fraction of a second every century, and, in the same period, the Moon recedes from the Earth by several meters. Traced backwards in time, it can be deduced that several billion years ago the Moon orbited just a few tens of thousands of kilometers away from the Earth, and a terrestrial day matched a lunar "month" of just a few hours; the Moon would have loomed apparently motionless over one hemisphere of the Earth, a great, hot orb that produced solar and lunar eclipses every few hours. Before this, so the fission theory goes, the Earth and Moon were physically joined.

There are some intriguing variations on the fission theme, including theories that incorporate Mars into the material of the original spinning proto-planet. One theory suggests that the Moon was a mere by-product of an Earth–Mars centrifugal separation – a smaller "droplet" of material that appeared between the two main planets. Another theory of the simultaneous formation of the Earth, Moon and Mars envisions a rapidly spinning twin-lobed proto-planet; the massive central hub became Earth, while the lobes were flung off to form the Moon and Mars. Needless to say, investigations into these theories are bound to encounter incredibly complex dynamics.

Although fission might explain why the Moon's core is not as massive as the Earth's and the similarity in their oxygen isotopes, support for the theory has almost completely evaporated among planetary scientists. One elaboration of the theory suggests that the roughly circular basin of the Pacific Ocean represents the scar of the Moon's violent birth, the Atlantic Ocean having been a split in the thin terrestrial crust to accommodate the huge mass of material torn from the other side of the planet: hence the matching Atlantic coastlines of Africa and South America. Current geophysical theories do not support this assertion – the Atlantic basin itself is a very young feature, just 130 million years old. Fission fails to explain why the Moon's orbit is inclined to the ecliptic by 5°; had the Moon been flung off from the Earth's equatorial bulge then it might be expected to have originally orbited above the Earth's equator, gradually being tugged by the Sun's gravity to assume a path in line with the ecliptic. The speed at which the original proto-planet must have been revolving in order to throw off a mass of material is so great – about 16,000 km/hour, giving the system twice the amount of angular momentum its present state – that astronomers consider it highly implausible.

Co-accretion

According to the theory of co-accretion, around 4.6 billion years ago, the Moon condensed from a primordial cloud of gas, dust and assorted debris in orbit around the Earth. Most of the Moon's mass may have accumulated into one single object in a period as short as 100 million years. Were the Earth and Moon sister planets, the relative abundances of elements such as iron ought to be similar; so too their overall densities. It has been suggested that the Earth may have accreted first out of denser material, leaving the Moon to sweep up any leftover lighter elements. Although co-accretion neatly explains the identical oxygen isotopes in lunar and terrestrial rocks, it fails to account for the lack of water and other

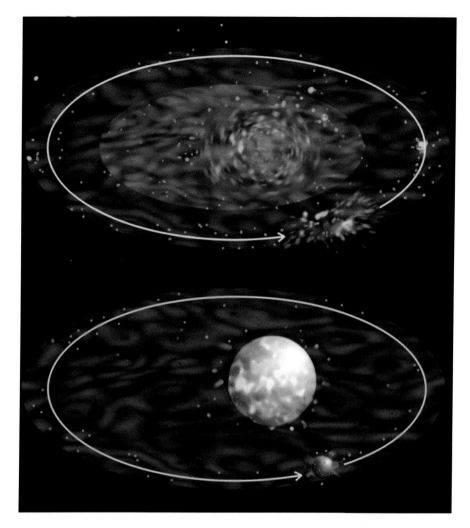

Fig. 1.2. Co-accretion model of the Moon's gradual formation from a cloud of debris in orbit around the embryonic Earth. Credit: Peter Grego

volatiles in the Moon's rocks: how could one of the driest worlds in the Solar System have formed alongside one of the Solar System's most watery ones? Moreover, the orbital dynamics of the Earth–Moon system do not square in the co-accretion theory – if the Moon had formed in close proximity to the Earth, its orbit would have initially been equatorial, later conforming to the plane of the ecliptic as a result of the gravitational tug of the Sun.

Capture

The tiny moons of Mars and many of the small satellites orbiting the four gas giants are undoubtedly captured asteroids. Those satellites that have highly inclined and retrograde orbits around their primaries are prime candidates for

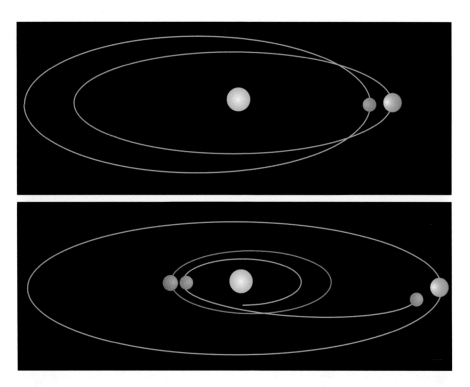

Fig. 1.3. Once an independent planet orbiting the Sun, the Moon was gravitationally captured by the Earth. One theory postulates that the original Moon had an orbit inclined 5° to the Sun and was captured when it approached within 64,000 km of the Earth. Another capture theory postulates that the Moon once orbited the Sun more closely than Mercury and was thrown out towards the Earth after Mercury and the Moon made a very close pass. Credit: Peter Grego

once having been independent objects. Could the Moon itself have once been an independent planet that was captured by the Earth's gravity?

In the capture scenario, gravitational perturbations caused the orbit of an inner planet to become so eccentric that it came within the range of capture by the Earth. It has been suggested that this planet's original orbit lay close to the Earth's, inclined to the plane of the Earth's orbit by about 5°. One model places capture at around 2.5 billion years ago, when the Moon approached within 64,000 km of the Earth. Another model proposes that the Moon originally had a circular orbit some 50 million kilometers from the Sun, placing it well inside Mercury's orbit. An exceptionally close encounter between Mercury and the Moon may have been sufficient to have both perturbed Mercury into its current eccentric orbit and propelled the Moon across the gulf of space into the Earth's gravitational embrace. The theory may explain the similar crater-sculpted faces of the Moon and Mercury, caused when both objects encountered asteroidal debris in the inner Solar System.

Wherever the planet Moon may have originated, if it had been an independent planet at one time, its momentum upon encountering the Earth needed to have been reduced in some dramatic fashion for it to have been forced to assume an orbit around the Earth; otherwise it would have flown by the Earth altogether. Some capture proponents postulate that the Moon ploughed into a dense barrage of large meteoroids and asteroids in Earth's orbit; this absorbed a substantial proportion of the Moon's orbital energy and slowed it in its tracks. This debris may have carved out some of the large lunar-impact scars – certainly, some of the ancient impact basins correspond in age with the proposed era of capture.

Capture theory may explain why the Moon now orbits the Earth at a 5° inclination to the ecliptic, in addition to accounting for the marked difference in the density and composition of the Earth and Moon. However, the theory does not explain the identical oxygen isotopes in Earth and Moon rocks, which imply that the Earth and Moon were formed in the same vicinity of the early Solar System.

Collision

Sometimes called the "Big Whack," this catastrophic hypothesis throws a sizable planet into the ancient Earth, causing a violent collision in which a vast cloud of debris was splashed into Earth's orbit. Much of this material eventually coalesced to form the Moon. The 'Big Whack' is currently the most favored theory of the Moon's origin because it solves more of the problems than any other theory.

Computerized studies have suggested that a Mars-sized planet (around half the diameter of the Earth) smashed into the young Earth with a glancing blow that disintegrated it and the Earth's mantle at the site of impact. The heavy iron-rich core of the impactor was largely absorbed into the core of the Earth, and a vapor-

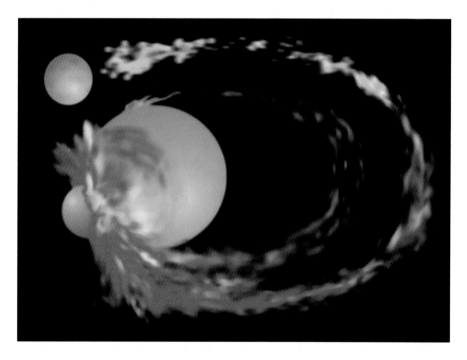

Fig. 1.4. The formation of the Moon is now widely thought to have been the result of a gigantic Mars-sized planetary impact with the embryonic Earth. The core of the impactor joined the body of the Earth, while the mantles of both objects were mixed and splashed out into space. Some of this material returned to the Earth, but a substantial proportion remained in orbit and gradually accreted to form the Moon. Credit: Peter Grego

ized stream of the mixed mantles of the Earth and the impactor was splashed into near-Earth orbit. This explains why the density of the Moon equates with that of the Earth's mantle, and why the Moon lacks a sizable iron core. The outer regions of this temporary ring of material (that lying beyond the Roche limit of gravitational disruption) quickly condensed into the Moon, perhaps at a distance of around 60,000 km.

If a massive impact had occurred, the extremely high temperatures produced would have been sufficient to boil much of the water and volatile material off into space, and that's why these substances are not abundant within the Moon's rocks. Mineralogical surveys conducted from lunar orbit support the "Big Whack" theory. The amount of iron in the Moon as a whole is significantly low. If the Moon had been a product of simple centrifugal fission or had been formed by co-accretion, the amount of iron would have to be much closer to that found in the Earth's mantle and crust.

Cosmic impact is becoming recognized as a major force in the evolution of the Solar System. Gone are the days when our own little corner of the Universe was viewed as a place of eternal tranquility, save for the occasional cometary visitor. The early Solar System saw countless major impacts, and the evidence can be readily observed in the impact scars of the terrestrial planets and the satellites. Planets far bigger than the Earth have been knocked askew by cosmic collision. A collision between the embryonic Mercury and Venus may account for their unusual properties – Mercury's huge iron core and Venus's slow, retrograde axial rotation. Gas giant Uranus lies on its side in space, so to speak, its axis inclined 98° to the ecliptic, possibly because a substantial object smashed into Uranus and knocked it on its side. So, the idea that a much smaller planetary collision brought about the Moon's birth has now gained great respectability among planetary scientists.

The Moon's Development

A deep layer of hot, molten rock is thought to have covered the Moon very early in its history. Conditions favoring such a hot young Moon are likely to have existed immediately after the "Big Whack," when the growing Moon's material was accreting so rapidly that huge quantities of heat were generated and maintained. Today we find large amounts of anorthosite in the Moon's highland crust; this is a mineral that forms exclusively in hot molten magma. Minerals like anorthosite would have been less dense than the magma melt, and they floated to the surface, aided by upwelling magmatic convection currents, to accumulate in gigantic "rockbergs" at the Moon's surface. These rafts of anorthositic (feldspar-rich) material became the first substantial crust of the Moon. Denser (iron- and magnesium-rich) minerals crystallizing in the melt sunk into the depths of the magma. This material was later to become the Moon's mantle, from which the relatively iron- and magnesium-rich lavas of the lunar maria were later derived. It is thought that the Moon's differentiated crust and mantle may have formed by about 4.4 billion years ago, only a couple of hundred million years after the Moon's formation.

During the next several hundred million years, further melting deep within the mantle caused pockets of hot magma to rise and intrude into the crustal material, some of it breaking to the surface and filling many of the young asteroidal impact basins. One component of many lunar rocks, called KREEP (K – potassium, Rare

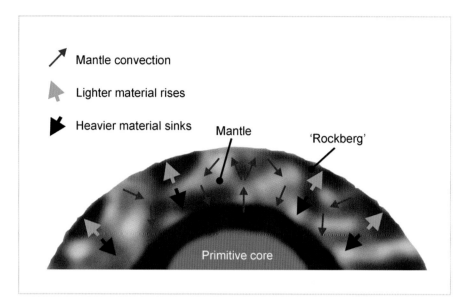

Fig. 1.5. A deep layer of hot molten rock covered the Moon early in its history. Lighter, less dense minerals like anorthosite would have floated to the surface of the magma melt, aided by upwelling convection currents, accumulating in gigantic "rockbergs," which became the first substantial crust of the Moon. Denser (iron- and magnesium-rich) minerals sank deep in the magma, later to become the Moon's mantle, from which the lavas of the lunar maria were later derived. The Moon's differentiated crust and mantle may have formed only a couple of hundred million years after the Moon's formation. Credit: Peter Grego

Earth Elements, and Phosphorus), represents the last vestiges of the Moon's sub-crustal magma ocean and was formed on the border of the crust and mantle. KREEP was brought to the surface mixed within magmatic intrusions and spread in lava flows, and it has been redistributed around the Moon by impacts. The unique chemical makeup of KREEP enables it to be used as a chemical tracer to chart the convoluted history of lunar volcanism and asteroidal bombardment.

In order to make sense of the history of the Moon's surface, lunar geologists use a relative time scale based on several major impact events – the Nectaris impact (3.9 billion years ago), the Imbrium impact (3.8 billion years ago), the Eratosthenes impact (3.2 billion years ago) and the Copernicus impact (1 billion years ago). The Pre-Nectarian Period spans the time between the Moon's formation, 4.6 billion to 3.9 billion years ago. Several major asteroidal impacts took place during this period, including, at around 4.2 billion years ago, the impact that created the South Pole–Aitken basin on the Moon's far side, so eroded that it is barely visible today. This was followed by the impacts that produced the basins of Australe, Tranquillitatis, Fecunditatis, Nubium, Smythii, Schiller-Zucchius and Apollo. In all, around 30 major asteroidal impacts occurred during the Pre-Nectarian Period.

The Nectarian Period, from 3.9 to 3.8 billion years ago, saw the formation of the basins of Nectaris, Moscoviense, Korolev, Humboldtianum, Humorum, Crisium and Serenitatis. The Imbrian Period, from 3.8 to 3.2 billion years ago, marks the last stages of major lunar bombardment, and encompasses three major events: the formation of the Imbrium, Schrödinger and Orientale basins. Lava flows spread across the broad circular expanse of the Imbrium basin, obliterating its inner

Fig. 1.6. A sample of KREEP taken from Mare Cognitum. KREEP represents the last vestiges of the Moon's hot magma ocean, and was distributed around the Moon's surface by volcanism and impact. KREEP is made up of Potassium (chemical symbol K), rare earth elements (REE) and phosphorus (P), and its unique chemical makeup allows it to be used as a marker for tracing the convoluted history of the Moon's surface. Credit: NASA/Johnson Space Center

structures, but Orientale's lava flooding was restricted to its center and just a few outlying areas. Orientale's appearance at the end of the major phase of lunar bombardment and its lack of lava flooding makes it is one of the Solar System's best-preserved major impact sites.

Volcanic activity transformed the appearance of all the near-side marial basins, while those of the far side were only partly filled with lava flows. Basaltic in composition, the marial lava flows had the viscosity of motor oil and were ductile and fast-moving. The dark seas visible with the naked eye were formed by layer upon layer of this runny material. More than 30% of the near-side was covered with mare material, compared with less than 3% of the far side. The discrepancy can be explained by the fact that the Moon's far-side crust averages twice the thickness of

Fig. 1.7. After its formation, the Moon was extensively bombarded by asteroids and meteoroids **1**: 4 billion years ago. A number of major impacts carved out the marial basins **2**: 3.8 billion years ago, which were subsequently flooded by lava **3**: 3.3 billion years ago. The Moon's face has appeared essentially the same since this period, except for a number of small crater-forming asteroidal impacts **4**: 100 million years ago. Credit: Peter Grego

the near-side crust, and since the Moon has been locked in a synchronous orbit since its earliest days, its Earth-turned hemisphere has always been under the greatest gravitational pull. The eastern maria of Tranquillitatis, Crisium and Fecunditatis were extruded around 3.5 billion years ago, preceding the in-filling of Imbrium and Procellarum in the western hemisphere by several hundred million years. The marial lava is thought to have formed in a hot radioactive decay melt layer, in places 100 km thick and up to 200 km beneath the lunar surface.

Beginning with the Eratosthenes impact around 3.2 billion years ago, the Eratosthenian Period saw the dwindling of large-scale lunar vulcanism and a co-incidental steep decline in the rates of major impacts. The Copernican Period began 1 billion years ago with the formation of one of the Moon's grandest craters, Copernicus. Most of the craters with bright ray systems were formed during this last period in lunar geological history, most notably Aristarchus and Tycho, formed around 300 and 100 million years ago respectively. The Copernican Period now shows the effects of surface erosion by innumerable meteoroids and micro-meteoroids – a relentless pummeling that has gradually rounded the lunar hills and produced a layer of fine soil.

Geologists consider the Moon to be an essentially dead world. The only major surface changes likely to take place in the next few billion years are going to be the result of asteroidal impact and/or human activity.

Inside the Moon

The upper few kilometers of the Moon's crust is known as the "megaregolith," and it consists of a layer of crushed and shattered rock produced during the era of intense asteroidal bombardment between 4.2 and 3.2 billion years ago. On average, the Moon's crust is 60 km thick on the near side and up to 100 km thick on the far side. Beneath this lies a solid mantle 800 km thick that (between 600 and 900 km beneath the lunar surface) produces moonquake foci. Moonquakes average less than a mild magnitude 2 on the Richter scale, and they have been found to occur in response to stresses in the crust, which are more prevalent when the Moon is closest to the Earth. An average of 3,000 moonquakes takes place each year – annually there are 300 times as many quakes of similar magnitude on the Earth. Lunar seismic disturbances also occur as a result of random meteoroidal impacts. Since there is no seismic activity on the Moon other than that produced by natural causes, seismometers operated on the Moon can measure disturbances several orders of magnitude fainter than those detected on the Earth. Seismic amplitudes of one millionth of a centimeter can be measured on the Moon. Seismic measure-ments have given tantalizing clues as to the internal construction of our satellite. The types of seismic waves detected include pressure waves and shear elastic waves which cannot travel through fluid, proving that the waves, had passed through a solid mantle. Long time delays in the arrival of seismic waves also suggest that the lunar crust is highly fractured and faulted.

There are some indications that there is a partial melt zone where the highly pressurized mantle can flow, some 1,200km beneath the Moon's surface. Beneath this layer, some constituents of the core, at a temperature of 1,000°C, are likely to exist in a molten state. Estimates of the diameter of the Moon's core are uncertain, and range from 500km for a dense iron-rich core to 1,500km if the core is less

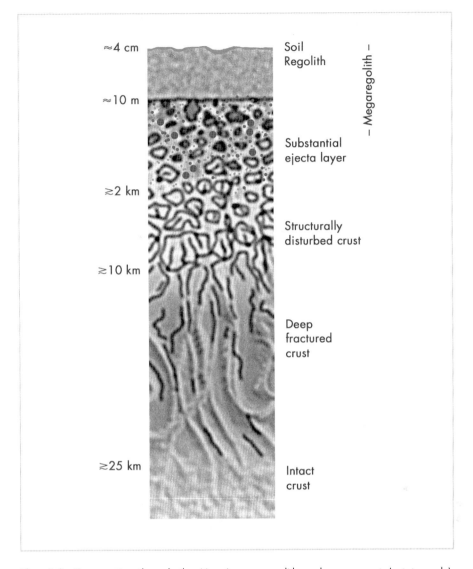

≈4 cm — Soil / Regolith

≈10 m

Substantial ejecta layer

≥2 km

Structurally disturbed crust

≥10 km

Deep fractured crust

≥25 km

Intact crust

— Megaregolith —

Fig. 1.8. Cross-section through the Moon's megaregolith and upper crust (not to scale).
Credit: Peter Grego

dense. The composition of the Moon's core is currently unknown, but if it is made up of iron and nickel like the Earth's core, it is odd that there is no trace of a global dipole magnetic field (a field with two poles), entities produced by dynamo effects.

On average, the lunar magnetic field measures about 38 nano Tesla (nT), a measurement of magnetic force) – the Earth's field at the equator is 1,500 times greater, measuring some 60,000 nT. Studies of remnant magnetism frozen into ancient lunar rocks as they solidified show that the lunar magnetic field strength in the past was at least 3,000 nT and was likely to have been produced in the hot molten interior by a dynamo effect similar to that which produces the Earth's magnetic field. The only appreciable magnetic fields now found on the Moon exist in

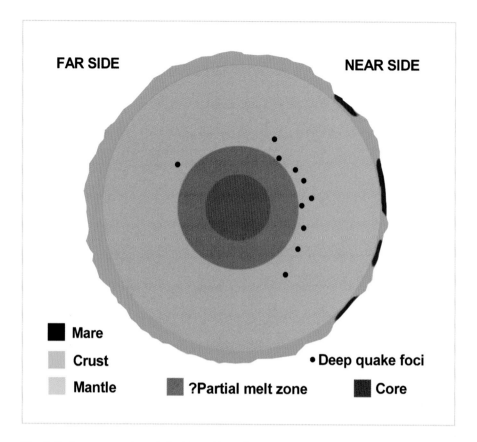

FAR SIDE

NEAR SIDE

- ■ Mare
- ■ Crust
- ■ Mantle
- ■ ?Partial melt zone
- • Deep quake foci
- ■ Core

Fig. 1.9. Cross-section through the Moon (the surface topography is not to scale). Maria (almost entirely on the near side) have average depths of several kilometers and overlie a crust several tens of kilometers thick. The highland crust is considerably thicker, having an average thickness of 70 km, and in places on the far side has a thickness exceeding 100 km. Beneath the crust lies the solid mantle, around 800 km thick. Beneath the mantle lies a probable partial melt zone, around 1,200 km beneath the Moon's surface. The boundary between the mantle and this Zone is the source of most moonquakes. The Moon's inner core is probably iron rich and smaller than 500 km in diameter. Credit: Peter Grego

localized pockets on the surface, notably in the vicinity of mysterious "swirls" such as Reiner Gamma (possibly the sites of cometary impacts) and, to a lesser extent, in some areas of the maria.

Gravity maps of the Moon show that certain well-defined areas on the Moon possess a considerably higher gravitational pull than do their surroundings. These patches of high-density material are called mascons (mass concentrations), and most of them are located at the centers of large, lava-flooded impact basins. There are more than a dozen notable mascons on the Moon's near side, the largest of which lie in Mare Imbrium, Mare Serenitatis, Mare Crisium, Mare Nubium and Mare Fecunditatis. Mascons were formed during the excavation of the near-side impact basins, when vast quantities of lunar crustal material were removed, leading to an immediate uplift of denser rocks from the Moon's mantle. This process, and the subsequent flooding of the marial basins by denser lava flows, led to the presence of greater concentrations of mass near the Moon's surface. Mascons indicate that the interior of the Moon must have been fairly rigid since the epoch of basin excavation; a hot ductile interior would not have been strong enough to bear the extra crustal mass for such a long period, and isostatic equilibrium would have resulted. Some areas on the Moon's far side also display positive gravitational anomalies, but they are not as pronounced or as clearly defined as on the near side. Mare Orientale, a large basin in the western hemisphere that straddles the near and far-sides of the Moon, displays the most unusual gravity profile: its center is strongly positive, but the region between its inner ring and outer ring is strongly negative. Negative anomalies are also found in the far-side basins of Hertzsprung, Mendeleev and Tsiolkovsky. Their presence indicates the greater thickness of the lunar crust on the far side, which prevented a post-impact uplift of denser mantle material, and caused the absence of significant flooding by lava.

Types of Lunar Rock

The Moon has never possessed an appreciable atmosphere, and its surface has been subjected to none of the processes of erosion found on the Earth. Extremely ancient rocks untouched by air or water are to be found all over the Moon's surface. The youngest lunar rocks – the basaltic lava that filled the near-side impact basins – date from around 3 billion years, which makes them about as old as the oldest rocks found on the Earth.

All of the loose material that overlies the solid rocky crust of the Moon is named the "regolith," and it includes the largest boulders to the finest granules of soil. Regolith consists largely of fragments of the local rock broken up by meteoroidal impacts – plus a small proportion of material that originated in large impacts further afield – overlain with a few centimeters of lunar soil. Lunar soil is comprised of rock fragments derived from the local material, various fragments of minerals, metals and glass agglutinates bound together by impact melt. In some locations, the soil contains tiny glass spheres, formed when high-viscosity lava was sprayed above the surface in fire fountains. This glass is composed of material that originated around 400 km beneath the surface, magma that climbed through channels in the lower mantle and crust relatively rapidly. Having experienced relatively little chemical interaction with the solid rocks surrounding it, the lunar volcanic glasses therefore represent the best samples of material from deep within the Moon.

Typical mare regolith consists of a layer of pulverized boulders, rocks and rubble from 2 to 8 meters deep, overlain with a few centimeters of soil. The regolith tends to be deeper in the highlands, ranging from 20 to 30 m in depth, but the regolith on the floors of some of the more recently formed large craters may be just a few centimeters deep. Large moonquakes and impact events shake the Moon's surface, causing the lunar soil to creep down slopes to accumulate in lower ground. Larger boulders can also be dislodged from time to time, and as they roll downhill they leave clear tracks in the soil.

Because there was never any substantial lunar atmosphere, and the surface was never covered with any expanses of water, not even the most primitive forms of life developed on the Moon. Thus, many rocks commonly found on the Earth – the product of atmosphere, water and life in combination with geological processes – have no lunar counterparts. The Moon has no clastic sedimentary rocks, such as conglomerate or sandstone, nor has it any biogenic sedimentary rocks, such as coal

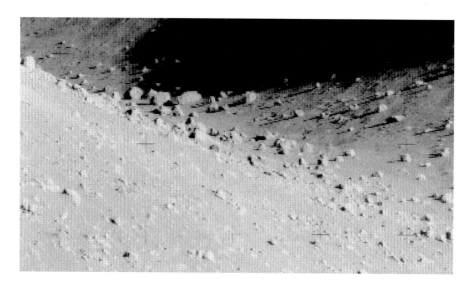

Fig. 1.10. The Moon's regolith is made up of fine-grained dust, mixed with rocks and larger boulders. This is a closeup view of the regolith in Hadley Rill, a lunar valley. The image shows a mixture of fine soil, rocks and substantial boulders that have rolled down the valley's slopes. Credit: NASA

or chalk. Because there has been no lunar tectonic activity or large-scale crustal movement, metamorphic rocks such as schist and gneiss, formed slowly through crustal pressure and heat, are not found on the Moon. All the Moon's rocks have been formed in processes that involved high temperatures; this includes igneous rocks that formed at high pressures deep beneath the Moon's crust, volcanic rocks that solidified at the Moon's surface and various shock-metamorphosed rocks and breccias that were created during the high temperatures and pressures of impact.

Although the Earth and Moon are strikingly different worlds, many of the Moon's rocks do have terrestrial counterparts, and geologists have few problems in recognizing the structure and mineral composition of lunar rocks when they are examined through the microscope. One notable difference is the complete absence of water in lunar rocks; all terrestrial rocks, even igneous rocks, contain a small amount of water; the extreme dryness of lunar rocks is a major clue to the Moon's origin, in that the high temperatures during the formation of the Moon boiled off the water and other volatile chemicals.

Basalt, anorthosite and breccia make up the three main types of lunar rock. Basalt is a dark-colored volcanic rock that once flowed as lava and filled all the marial basins of the near side and some areas on the far side. Around 30% of the near side surface is basaltic, while just 2% of the surface of the far side is composed of this rock type. The basaltic lavas did not all originate at the same level within the Moon's mantle – pockets of magma at different depths assumed variations in chemistry that developed over time. Typically around 50% of lunar basalt comprises the mineral pyroxene, a calcium–magnesium–iron silicate. Its grain size averages less than 1mm, an indication of its rapid cooling after extrusion. Lunar basalt closely resembles the basaltic material that makes up the Earth's oceanic crust; while the lunar basalts were extruded onto the Moon's surface more than 3 billion years ago, the Earth's oldest oceanic crust is less than 200 million years old.

Fig. 1.11. A basaltic rock taken from Mare Tranquillitatis. The lunar maria are all composed of this dark type of rock, which flowed onto the Moon's surface more than 3 billion years ago. Credit: NASA/Johnson Space Center

The Moon's bright highland regions – the shattered, impact-battered remnants of its original crust – are made up of rocks rich in the mineral feldspar, and are generally termed "anorthositic" rocks. Anorthosites are made up of 90% or more of the mineral plagioclase, a very light-colored form of the mineral feldspar, and crystallized out of the Moon's early magma "ocean" around 4.5 billion years ago to form the first lunar crust. Other plagioclase-rich rocks make up a group called the "Mg-suite," whose pyroxene and olivine minerals are rich in the metallic element magnesium (Mg). The most abundant Mg-suite rocks are norite (plagioclase–pyroxene rock) and troctolite (plagioclase–olivine rock). The Mg-suite rocks date from around 4.5 to 4.2 billion years, and were formed by a process of assimilation in numerous magmatic intrusions into the anorthositic crust. Mg-suite rocks probably make up a major constituent of the Moon's lower crust, as they are found in abundance in the ejecta from the major impact basins – impacts large enough to have excavated material from deep within the crust.

Breccia is a composite rock made up of fragments created through the processes of impact shattering, mixing, melting and recrystallization of rock during the high energies released during meteoroid and asteroid impacts. Typical breccias contain coarse fragments embedded in a fine-grained crystal matrix. Breccias are described as monomict if they contain fragments of only one rock type; most breccias are polymict and are composed of more than one type of rock.

All three rock types – mare basalt, anorthositic material and breccia – can be found in any location on the Moon, though the proportions will vary, depending on whether the site is located in the highlands or a mare. An average of 90% of the

Fig. 1.12. Anorthosite makes up the Moon's highland crust, and is the oldest rock type found on the Moon. This sample is 4.19 billion years old and was taken from the Descartes highlands. Credit: NASA/Johnson Space Center

material gathered at any one lunar location will be representative of the underlying bedrock, though it has been mixed, deformed and melted during a multitude of impact events. A small proportion of material found in any spot will come from much further afield, debris thrown out by distant impacts hundreds of kilometers away.

Shaping the Moon's Surface

Impact

Much of the Moon has been intensely sculpted by meteoroidal and asteroidal impact. All of the terrestrial planets – Mercury, Venus, Mars and the Earth – are thought to have been subjected to similar levels of bombardment, but only Mercury's surface resembles that of the Moon. Through the eons, the substantial atmospheres of Venus, the Earth and Mars have served as an effective impact buffer, allowing only a few of the biggest incoming objects to wreak large-scale damage. Atmosphereless Mercury and the Moon have been fully exposed to the harsh vacuum of space and subjected to bombardment by interplanetary dust, meteoroids, asteroids and comets, in addition to x-rays, gamma rays and cosmic rays.

Fig. 1.13. A breccia taken from the valley of Taurus Littrow. Lunar breccias are made up of fused fragments of material derived from impacts. This sample consists of fragments of glass, minerals and rock cemented together in a glassy matrix. Credit: NASA/Johnson Space Center

The Moon's surface bears stark testimony to eons of impact. It has been estimated that around 3 trillion craters over a meter in diameter dot the lunar surface. The Moon's near side displays around 300,000 craters larger than a kilometer, no fewer than 234 of these being larger than 100km across. Large tracts of the lunar surface have been modified by volcanic activity and other geological processes, including faulting. The lunar crust has never been subjected to any appreciable tectonic activity, and large-impact features many billions of years old can be clearly traced; some of them were actually formed before life on Earth appeared. The overwhelming majority of lunar craters display the hallmarks of impact formation. Leaving aside the evidence of the lunar rocks themselves, there is a clear pattern of impact-crater morphology spanning all size ranges, from the tiniest micrometeorite impact pits to the vast asteroidal impact basins. The observed morphology of lunar craters perfectly matches computer studies of impacts, in addition to ballistic impact experiments performed in laboratories and field studies of terrestrial craters (both natural impact features and manmade explosion craters). Many of the lunar rocks themselves could only have been formed as a result of the sudden high temperatures and pressures produced during impact events. There is no evidence to support the idea that any of the larger craters represent the rims of ancient lunar volcanoes, nor is there any evidence to suggest that any major craters were formed by violent crustal explosions or as a result of crustal

Fig. 1.14. A small number (just a few dozen) of meteorites found on Earth are known to be material that originated on the Moon, having been thrown from the lunar surface by the explosive power of impacts. Only 0.08% of catalogued meteorites are known to be lunar. This rare material may have been blasted from around 20 individual impact events during the past 10 million years. These tiny fragments of lunar-highland breccia and mare basalt, owned by the author, are chips from the catalogued lunar meteorites. Credit: Peter Grego

collapse after magmatic subsidence. If some craters did happen to be volcanic in origin, then a number of these features, frozen in various stages of formation, would likely exist, but this has not been not observed.

The lunar surface is continually pelted with a rain of micrometeoroids (a term applied to meteoroids smaller than 1 mm in diameter), with an average rate of one micrometeoroid impact per week per square meter. Though small, micrometeoroids impact at very high velocities, ranging from a minimum of 2.4 km/second (the Moon's escape velocity) to 40 km/second. Since the Moon has no atmosphere to slow these objects as they plummet to the surface, the average impactor hits with an unimpeded interplanetary velocity of more than 70,000 km/hour. A micrometeoroid that would burn up completely in the Earth's atmosphere will hit the lunar surface and produce a tiny impact crater. The surfaces of lunar rocks studied under the microscope are indented with countless tiny impact pits, many of them surrounded by a splash pattern of melted material. Rugged alpine landscapes cannot be found anywhere on the Moon. Wearing down solid rock, the continual rain of meteoroids has served to erode the Moon's surface, smoothing mountain peaks that were once as sharp and jagged as their terrestrial counterparts.

Impact craters can measure more than 50 times the diameter of the impactor, and the volume of material that they excavate can be hundreds of times the

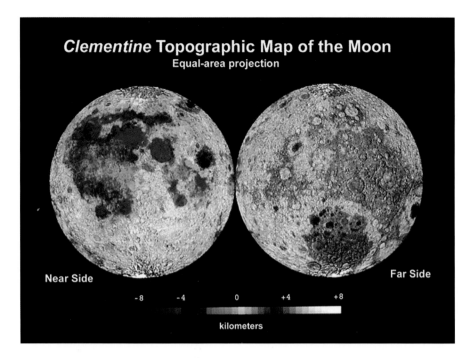

Fig. 1.15. Topographic map of the Moon (Clementine laser altimetry: red = high, purple = low) showing both lunar hemispheres. The giant (2,500-km-diameter) South Pole–Aitken Basin can be easily discerned on the lunar far side. Credit: The Clementine Project/USGS

impactor's volume. Only the largest meteoroids are capable of penetrating the regolith to the solid crust beneath. A meteoroid about 10 m across would be sufficient to penetrate the typically deep regolith in the highlands. Any meteoroid large enough to slice down into the lunar crust generates tremendous pressures and temperatures as its kinetic energy (the product of the meteoroid's mass and the square of its velocity) is converted into shock waves and heat that is imparted into the surrounding crust. The crust beneath the impactor is compressed and the surrounding material is pushed downwards and outwards. An ultrahot bubble of expanding molten material with a temperature of several million degrees is formed as the impactor and the surrounding rocks are nearly instantaneously vaporized. The edge of the crater is deformed and uplifted as a plume of excavated material, made up of vaporized rock and larger rock fragments, is blasted outwards from the impact site. As the crust decompresses, rebound effects produce a central uplift in larger craters, and a substantial layer of melted rock accumulates within the crater's bowl.

The excavated material is distributed around the crater in an ejecta blanket. Some larger fragments may be hastily welded together in breccias, while smaller droplets of material may cool and solidify in flight, forming tiny glassy beads or larger "bombs." The ejecta itself is deposited in an ordered manner. The first material to be ejected comprises material that was close to the focus of impact near the surface, and this high-velocity material is launched steeply above the surface to be deposited at great distances from the crater. As the impact progresses, deeper material is excavated, but the overall energy of the impact dissipates. With progressively

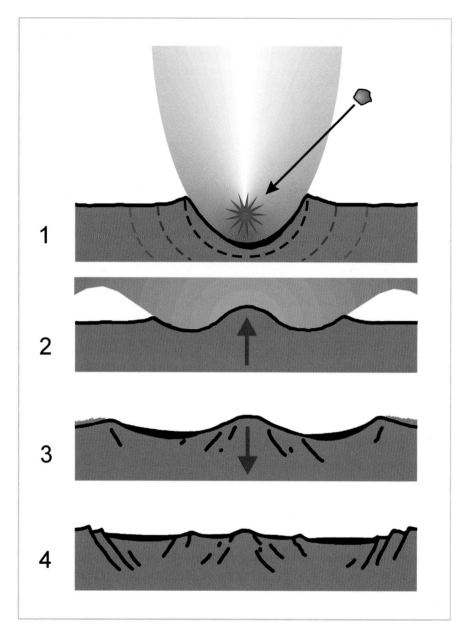

Fig. 1.16. Stages in the formation of a large impact crater. **1** Asteroid impact, the focus of an explosion deep inside the megaregolith. The crust is compressed and shockwaves propagate outwards, fracturing the crust. A plume of ejecta is thrown up from the Moon's surface as the edge of the explosion cavity is deformed and uplifted. **2** Immediately following the impact, the compressed crust rebounds, forming a central elevation. The crater enlarges as the ejecta sheet is thrown across the crater's surroundings. The material originally nearer the surface is ejected furthest, while deeper bedrock is lobbed shorter distances from the rim. **3** The uplift collapses, impact-melted rock and breccia collects in the crater. **4** The feature settles and collapses along points of crustal weakness, giving rise to slumping and internal terracing of the walls. Credit: Peter Grego

slower velocities, the ejecta is distributed ever closer to the crater, and the deepest excavated bedrock may barely be lobbed over the crater's rim. This produces an inversion of the ejecta blanket's layering compared with the original layering of the impact site; material once forming the top layers at the impact site are overlain by material that originally lay beneath it.

Craters smaller than about 10 km across are usually smooth, bowl-shaped depressions with rounded, raised rims, and they have an average depth of about one-tenth their diameter. They have simple ejecta collars, usually forming an area of rough terrain that extends to about a crater diameter away from the rim, beyond which can be found small secondary craters formed by the impact of larger fragments of debris excavated by the primary impact. Such simple craters can be seen dotted all over the maria and on the floors of large, flooded craters. Larger-impact features generally have a smaller depth to diameter ratio and they display progressively more complex forms. They often have sharp rims and their floors are covered with mounds of debris that has slipped off the steep inner walls, giving the crater rim a distinctly scalloped or polygonal outline. Their surroundings are mottled with a peculiar "herringbone texture" made up of a mass of V-shaped hummocks that point towards the crater. Some of this structure may be caused by elevations in the surrounding terrain that leave a "bow wave" pattern free of ejected debris.

Complex craters more than 20 km in diameter (see Table 1.1) often have broad, flattish floors with prominent central elevations. Beyond the rim there is a considerable amount of physical structure, including prominent radial ridges and grooves, in addition to secondary cratering caused by the impact of substantial chunks of excavated bedrock. Secondary craters are simple mechanical-impact pits formed at a far lower velocity than the original impact. The amount of material excavated by these secondary impacts can actually exceed the volume of material thrown out during the original high-velocity impact, since low-velocity impactors tend to be more efficient excavators than high-energy ones, which generate vast quantities of excess heat. Streams of ejected debris can produce lines of craters arrayed radially around their parent crater. These features can consist of an unconnected series of craters of roughly similar size (typically smaller than about one-twentieth the diameter of the parent crater), or in an interconnected chain of craters (some of them highly elongated) that runs a considerable distance across the surface. Some secondary-impact crater chains are so tightly knit that they resemble rilles, and can be mistaken for such at first glance. Such are the dynamics of impact that these radial structures do not follow perfectly straight lines – secondary-chain craters can follow quite curved, even sinuous, paths over the surface. The best place to view the entire array of secondary-impact structures is around the mighty crater Copernicus, since the structures were carved out of generally rather level mare material only comparatively recently in lunar geological terms.

Many large young craters are surrounded by bright ray systems. Rays occur where the regolith has been churned up by secondary impacts, exposing fresh material at the surface. Some rays actually have a distinctly different composition from the regolith that they overlie. Like fingerprints, no two lunar ray systems appear to be exactly alike. Some rays, like those around the little crater Linné in Mare Serenitatis, form near-circular halos. Others, like those around Kepler in Oceanus Procellarum, form a neat splash pattern that is equally bright all over. Other ray systems are made up of a network of bright individual radial rays that

Fig. 1.17a. A micrometeorite impact pit in a tiny glassy bead sampled from the lunar soil. The minuscule crater measures just 30 microns in diameter. Credit: NASA.
b. Palmetto. A close-up of impact cratering in the Descartes highlands. The large crater at top center is Palmetto, 1 km in diameter, but craters as small as 1 meter can be resolved in the image. Palmetto is deep, bowl-shaped and has rounded walls. Credit: NASA. **c.** Linné crater in Mare Serenitatis, 2.5 km in diameter. Linne is deep and bowl-shaped, but has a sharp rim and a bright collar of ejecta. Credit: NASA. **d.** Lalande crater, in the highlands south of Mare Insularum. The crater, 24 km in diameter, has a sharp rim, slumped internal walls and traces of impact-melt floor from which protrudes a small central peak. Credit: NASA

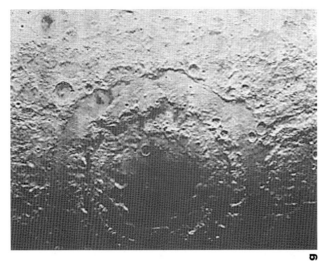

Fig. 1.17e. Tycho is a large, prominent impact crater in the Moon's southern uplands, 85 km across. It has a sharp rim, wide internal walls with prominent terracing and a large group of central peaks that rise above a flattish floor filled with breccia and impact melt. Small secondary craters and a brilliant ray system emanate from Tycho. Credit: NASA. **f.** Schrödinger, an impact basin on the Moon's far side, measures 320 km in diameter. It has broad terraced walls and an impact-melt floor with a substantial, near-complete inner ring of mountains, but no central elevations. Credit: NASA. **g.** The Orientale basin is one of the Moon's largest and youngest major impact basins, with an outer ring measuring 930 km in diameter. It has multiple internal mountain rings of 620 and 400 km in diameter, and a flooded central area (Mare Orientale itself). A substantial system of secondary-impact craters, many of which are arranged in radial chains, radiates around the Orientale basin. Credit: NASA

Table 1.1. Some large complex lunar craters (youngest to oldest)

Feature	Diameter (km)	Depth (m)	Floor	Central elevation	Walls	Outer structure	Ray system	Age (y)
Aristarchus	40	3,630	Small, (20 km) with impact melt	Small single mountain	Broad, steep and terraced, with dark radial banding	Marked polygonal outline. Radial ridges and furrows. Dark impact melt collar.	Prominent rays extend up to 300 km	30 my
Tycho	85	4,850	Broad, slightly knobbly, with impact melt	Two main mountain blocks, the largest of which rises to 1,000 m	Strongly terraced	Some concentric structure, plus radial ridges and furrows. Dark impact melt collar.	Secondary crater chains absent. Has the most prominent lunar ray system, rays up to 1,300 km long	100 my
Copernicus	93	3,760	Broad floor, generally rougher in the south, with impact melt	A complex of mountains rising to 1,200 m	Broad and highly terraced	Large concentric structure around rim, with prominent radial ridges, furrows, and substantial secondary impact craters and prominent crater chains	Large "starburst" ray system with components up to 800 km long	900 my
Aristillus	55	3,650	Broad, flat impact melt	A dozen small central peaks	Broad and highly terraced	Prominent radial ridges and furrows, some secondary impact craters	Very faintly traceable	?2 by
Vieta	87	2,000	Broad, eroded and cratered	None	Eroded, some craters	Largely obliterated	None	?3.5 by

maintain the same width over their entire length; Tycho, in the southern uplands, displays such a system. Some bright lunar rays spread in only one direction, either in a broad fan of material or in a number of searchlight-like lines, like the twin searchlight-like beams emanating from the crater Messier A in Mare Fecunditatis. Experiments have shown that such asymmetric rays are produced by high-velocity impactors that approach the Moon's surface at angles shallower than 7°. The ray systems of numerous large craters appear to commence a short distance away from the crater rim, the crater itself being surrounded by a dark collar of material. The dark material is where low-velocity impact melt, composed of dark glassy material, splashed onto the crater's immediate surroundings. Erosion by meteoroids gradually churned up the ray material on the surface, fading it over time. Rays may only last a couple of billion years before being blended in with their surroundings.

Craters larger than 100 km in diameter are known as ringed basins. Those of the Moon's ringed basins that range from 140 to 175 km in diameter have a sizable central massif, caused by crustal rebound and uplift, surrounded by an inner ring of scattered mountains. Larger basins (up to 400 km in diameter) have a well-developed inner mountain ring but lack a central elevation. The largest impact structures found on the Moon are the multiringed basins. These vast scars of

Table 1.2. Some major lunar impact basins (youngest to oldest)

Basin	Outer ring (km)	Inner rings (km)	Obvious flooding	Age
Orientale	930	620/480/320	Inner ring	Imb
Schrödinger	320	150	Partial	Imb
Imbrium[a]	*1200*[b]	**670**	**Total**	**Imb**
Bailly	**300**	**150**	**None**	**Nect**
Hertzsprung	570	410/265	Inner ring	Nect
Serenitatis	**740**	**420**	**Total**	**Nect**
Crisium	**1000**	**635/500/380**	**Total**	**Nect**
Humorum	**820**	**440/325**	**Inner ring**	**Nect**
Humboldtianum	**600**	**275**	**Inner ring**	**Nect**
Mendeleev	330	140	None	Nect
Korolev	440	220	None	Nect
Moscoviense	445	210	Partial	Nect
Nectaris	**860**	**600/450/350**	**Inner ring**	**Nect**
Apollo	500	250	Partial	Pre-Nect
Grimaldi	**430**	**230**	**Inner ring**	**Pre-Nect**
Birkhoff	330	150	None	Pre-Nect
Planck	325	175	None	Pre-Nect
Schiller-Zucchius	**325**	**165**	**Partial**	**Pre-Nect**
Lorentz	360	185	None	Pre-Nect
Smythii	**840**	**360**	**Inner ring**	**Pre-Nect**
Poincaré	340	175	Partial	Pre-Nect
Ingenii	560	325	Inner ring	Pre-Nect
Nubium	**650**		**Total**	**Pre-Nect**
Fecunditatis	**650**		**Total**	**Pre-Nect**
Tranquillitatis	**775**		**Total**	**Pre-Nect**
Australe	**880**	**550**	**Partial**	**Pre-Nect**
South Pole–Aitken	*2500*		None	Pre-Nect

[a] Bold text indicates features located on the near-side.
[b] Figures in italics indicate buried, eroded, irregular or partially visible features.

asteroidal bombardment measure more than 400 km across, and contain multiple concentric mountain rings. The rings may represent frozen waves of crustal material caused by the collapse of the original central uplift immediately after impact, and the outer rings of multiringed impact basins may be far larger than the original asteroid impact crater. Examples of the variation in the morphology of impact features of different sizes can best be seen on the far side, where there has been little overt volcanic activity and lava flooding to alter their appearance (see Table 1.2).

Basin formation imparted tremendously powerful seismic waves into the lunar crust, which traveled around the Moon to converge on the other side, immediately opposite the impact site. The concentration of these seismic waves led to the formation of localized areas of chaotic terrain, where pre-existing structures were shaken to their foundations, producing a knobbly, textured landscape. This peculiar kind of topography can be found in and around the far-side crater Van de Graaf, a sizable basin 233 km across whose walls were turned into piles of rubble by the seismic waves generated by the formation of the Imbrium basin 3.8 billion years ago. The Orientale impact, just past the southwestern lunar limb, appears to have created a similar topographic disturbance in the region south of the crater Hubble, visible near the Moon's northeastern limb.

Basin Flooding

Welling up from the mantle, magmatic intrusions through weak points in the lunar crust have extruded onto the surface as basaltic lava flows. This material has flooded most of the near-side ringed basins to create the maria, prominent dark plains that cover a total of about 30% of the Moon's near side. Although they occupy substantial areas of the near side, the basaltic mare lavas actually make up less than 1% of the total volume of the lunar crust. The mare lava erupted from about 4 to 3 billion years ago. Samples of mare basalt melted in the laboratory take on the consistency of motor oil, an extremely ductile substance that is capable of

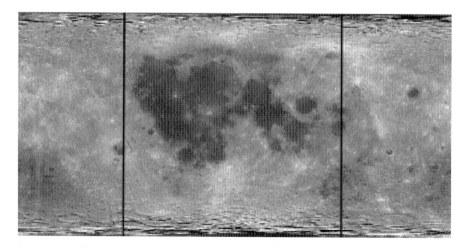

Fig. 1.18. Projection of the entire lunar surface. The central area bounded by lines is the Moon's near side. Note the remarkable difference in the number of large, dark, lava-filled basins between the two lunar hemispheres. Credit: The Clementine Project/USGS

spreading over large areas before cooling to a halt. The maria were not flooded in one single episode – each mare is made up of a number of different layers of lava flows, and distinct strata can be seen in the walls of many small craters in the maria. Different phases of lava flooding are quite evident telescopically, as many maria take on a patchwork appearance that suggests numerous phases of volcanic activity that erupted batches of lavas of slightly different chemical composition. Most of the marial lava flows were extruded in very large eruptions from extended volcanic vents dotted around the floors of ringed basins and larger impact craters. Masses of material flowed rapidly across the surface in broad sheets that traveled for hundreds of kilometers.

Cryptomaria

The Moon's surface appears to comprise two basic landscapes, dark-colored, lava-filled plains and bright-colored cratered highlands. Digging a little deeper will reveal a more complicated story regarding the lunar surface. Some small, relatively recently formed impact craters dotted around the Moon's highland plains are surrounded by collars of dark ejecta. Since spectral analysis shows that these dark halos consist of mare basalt, these dark-halod impact craters provide evidence for ancient buried lava plains known as cryptomaria. The dark-halod craters have excavated mare material that was originally extruded onto the lunar surface very early in the Moon's history, long before the mare basalts of the currently visible lunar seas were formed. The ancient basaltic lava flows of the cryptomaria were widely spread across large patches of the Moon's surface but were completely obscured by piles of debris, many tens of meters thick, thrown out by subsequent major impact events. The total area once occupied by cryptomaria may have amounted to as much as half that occupied by the visible maria. The area around Mare Crisium, Mare Marginis and Mare Smythii, extending to the Lomonosov-Fleming region just past the eastern limb, is dotted with dozens of small dark-halod craters, indicating the presence of deeply buried lava flows. Cryptomaria also lie in and around the flooded basin Schickard, near the Moon's southwestern limb, and to the west of Mare Humorum.

Lunar Volcanoes

As large-scale vulcanism dwindled, smaller volcanic vents gave rise to a variety of different features. Low, rounded hills called "domes" – typically a few hundred meters high and with bases several kilometers across – are the remnants of ancient lunar volcanoes, and their tiny summit craterlets represent volcanic vents. Even with such fluid lunar lava flows, domes built up when a balance had been struck between the supply of lava being erupted and the rate of cooling of the extruded material. West of Copernicus can be found the Moon's best location for volcanic domes; a group of six lie north of the crater Hortensius and another group of around a dozen lies north of the crater Milichius. Intermittent eruptions of volcanic ash and pyroclastic deposits produced steeper-sided volcanic cones. Low-viscosity lava, squeezed through very narrow volcanic vents, can produce fire fountains that spray a multitude of tiny droplets – successive eruptions are capable of building up quite steep-sided volcanic cones. It is likely that chunks of material

Fig. 1.19. Formed by fast-flowing lava, Sinuous rilles originate at volcanic vents and meander downslope. Lunar domes are squat features that were built up by lava and ash deposition when the moon was volcanically active. Many display summit craters (volcanic vents).

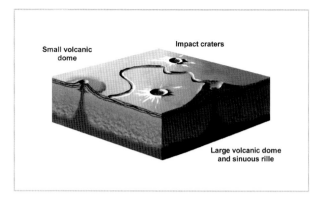

were torn away from the walls of the magmatic channel feeding the volcanic vents, and were spat out amid the fire fountain eruptions – examples of such "xenoliths" (xeno: stranger, lith: rock) may be found in the deposits surrounding the lunar cinder cones. The Marius hills in Oceanus Procellarum create one of the Moon's most spectacular volcanic landscapes: no fewer than a hundred volcanic cones and domes occupying an area of around 40,000 sq km.

Wrinkle Ridges

All the lunar maria display peculiar low winding ridges known as "wrinkle ridges." These features, technically termed "dorsa," average just a few tens of meters in height and a few kilometers in width, and as such are visible only under a very low angle of illumination when they cast shadows onto their surroundings. Dorsa are the same color as the terrain around them, and they cannot be discerned under a high Sun. There are two basic types of dorsa. The broader ridges have rounded, shallow-sloping margins and rise at an average angle of about 3° from the mare level. Narrower dorsa with slopes of about 15° can often be seen lying parallel to, or even superimposed upon, the broad ridges. Many of the major dorsa appear to follow the outlines of the mare basins. Some of them – those in Mare Imbrium, for example – seem to indicate the location of the basin's inner rings, buried beneath hundreds of meters of lava flows. In some cases – notably in Mare Imbrium – the highest points of the original inner ring structure actually protrude through the wrinkle ridges to stand proudly above the surface as isolated peaks. Some mare wrinkles appear to indicate the presence of ancient, completely buried craters, such as Lamont in Mare Tranquillitatis. Other wrinkles lie near the mare borders and appear to extend from semicircular bays, marking the presence of a crater's submerged wall and, in some cases – Letronne on the border of Oceanus Procellarum, for example – its central elevation. These features are likely to have formed when the surfaces of the newly formed maria contracted and compressed, buckling the surface. A number of mare ridges are the remnants of ancient lava flows; numerous examples can be viewed in Mare Imbrium. These features, which often have broad, ill-defined points of origin, have very low profile and their margins often extend into multiple well-defined lobes.

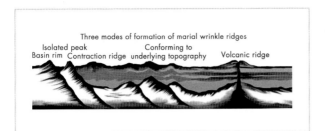

Fig. 1.20. Cross-section through a lunar mare and its mountain border. Wrinkle ridges have a number of possible modes of formation, including mare contraction, topographic moulding to underlying features and vulcanism.

Sinuous Rilles

Some active volcanic vents were the powerhouse behind the formation of sinuous rilles – narrow, winding valleys that superficially resemble dried-up terrestrial river valleys ("graben," German for "ditch"). Very fast flowing lava extruded from volcanic vents rapidly cut through the recently deposited mare lava flows. The vents are often circular and crater like with flat floors, sunk into the landscape without any appreciably raised rim. The rilles themselves are similarly sunk into the landscape, U-shaped in cross-section, with inner walls that slope by about 20°–30°. Having been formed by very runny lava, the channels always progress downhill, gradually diminishing in depth and width, although the overall slope of the landscape is not always evident. Some sinuous rilles appear to cut across higher ground, in apparent defiance of gravity. However, this higher ground is more than likely to have arisen by crustal buckling or arching some time after the sinuous rilles had been cut. It is likely that narrower, shallower sections of some sinuous rilles were originally roofed over, and the lava flowed underneath the surface in tunnels or "lava tubes" – volcanic features observed in many terrestrial locations, but on far smaller scales. Their courses were later exposed when the fragile tunnel roofs collapsed, and the remnants of the original roof can be observed on the rille floor, some large house-sized fragments having survived the fall intact. Unlike terrestrial river valleys, sinuous rilles do not have any outwash plains. Many beautiful examples of sinuous rilles can be found in the near-side maria and within some lava-filled craters. Vallis Schröteri (Schröter's Valley), the largest lunar sinuous rille, lies on the Aristarchus plateau in Oceanus Procellarum. Vallis Schröteri is more than 150 km long, averages 4–6 km wide and measures up to 1,000 m deep in places. Vallis Schröteri is of exceptional interest, since a smaller sinuous rille actually winds along the length of the main valley floor, formed when the source of lavas was dwindling. Nearby, the dozen or more components of the Rimae Aristarchus makes one of the finest groups of sinuous rilles.

Faults and Linear Rilles

Lunar tectonic activity ceased some time after mare flooding, around 3 billion years ago, and seismic studies show that the Moon's crust does not currently experience any appreciable movement. There is much evidence that areas of the Moon's crust were once under considerable tectonic strain, forces of tension and compression having produced breaks in the rock, or "faults." In addition to tectonic activity, faulting took place in and around major impact structures as

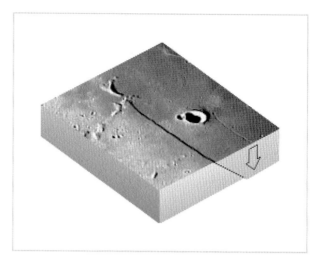

Fig. 1.21. The best example of a normal fault on the Moon is Rupes Recta in Mare Nubium. Crustal tension has caused the crust to crack, and the western block of crust has slipped down, leaving a prominent escarpment some 110 km long. Credit: Peter Grego

sections of the crust fractured and slumped in response to unbearable loads set up in the crust.

Normal faults are the most commonly observed types of fault on the Moon, and result from crustal tension; as the crust is pulled apart, the rocks may deform to a certain degree, but there comes a point when the crust cracks. Fault planes are usually inclined and rarely vertical. One side of the fault slips down in relation to the other, producing a freshly exposed rock face called a fault-scarp. Easily the best example of a normal fault on the Moon is Rupes Recta (often called "the 'Straight Wall'") in southeastern Mare Nubium. At first view, Rupes Recta appears to be an impressively steep wall dividing a 126 km section of the mare, but in reality the scarp face has a gentle gradient of 7°–30° and a height of around 500 m. Other clearly visible examples of normal faults can be found in the maria, where they show up well under low angles of illumination. Faults that have been produced in the lunar highlands tend not to have the clean-cut form of Rupes Recta. Rupes Altai is a normal fault that lies several hundred kilometers due east of the Straight Wall. Four hundred and eighty kilometers long, and somewhat sinuous along much of its length, Rupes Altai lies parallel to the southwestern border of Mare Nectaris, and actually marks part of the Nectaris basin's outer ring.

When a block of crust lying between two parallel normal faults subsides, a feature known as a graben rille results. Such valleys are rather common on the Moon. Because their origins are deep-seated, graben rilles can cut cleanly through pre-existing hills, mountains and craters without altering their course. By far the biggest lunar graben (far too large to be termed a rille) is Vallis Alpes, a spectacular rift valley that cuts neatly across the lunar Alps. Vallis Alpes is 180 km long and averages 10 km wide, with steep inner walls. Its flat lava-flooded floor, sunk about 1,000 m below the surrounding uplands, has a small lava-cut sinuous rille running down the middle. Many prominent graben cut across lunar maria, parallel to mare borders; these are known as arcuate rilles because of their curving paths. Prominent arcuate rilles lie near the southwestern border of Mare Tranquillitatis and the western and southeastern border of Mare Humorum. Intricate systems of interconnected linear rilles can be found in some maria and within many larger craters. Good examples include a beautiful set of rilles to the east of Triesnecker in Sinus Medii, those on the floors of Lacus Mortis and within the craters Gassendi

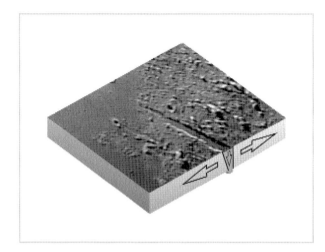

Fig. 1.22. Vallis Alpes, a large rift valley in the lunar Alps, was caused when crustal tension caused two parallel faults to appear across the mountains; the crust between the faults subsequently dropped down. This kind of feature is known as a graben, and many of the Moon's smaller linear and arcuate rilles were caused in the same manner. Credit: Peter Grego

and Posidonius. One of the more unusual of the Moon's rilles is the Hyginus rille in Sinus Medii, a valley 220 km in length that is punctuated by a series of small craters, including Hyginus itself. This may be a rare example of a hybrid rille – one that originally formed as a result of crustal tension but that later experienced volcanic activity along its length.

Dark-halo craters are also found to lie along some linear rilles. Unlike the dark halo craters that reveal the Moon's cryptomaria, these features are undoubtedly volcanic vents that developed along deep-seated rilles and pumped out clouds of volcanic debris – glass beads formed in fountains of low-viscosity lava, layers of ash and pyroclastic deposits. The best examples of these volcanic dark-halo craters can be observed on the floor of the large crater Alphonsus near the center of the Moon's near side; here, half a dozen sizable dark patches surround small craters that sit along rilles.

Mountains of the Moon

The Earth's crust is a patchwork of relatively thin plates of solid rock floating on a hot, dynamic mantle driven by convection currents. New crustal material is supplied by the mantle as active volcanic ridges push oceanic plates apart, causing them to collide with continental crust at its margins. Volcanoes, such as those in the "ring of fire" around the Pacific Ocean, continually deposit fresh masses of material on the surface. Collisions between the Earth's plates also builds vast mountain ranges; the Earth's largest mountain range, the Himalayas, resulted from a collision between the Indian and Asian continental plates that began about 15 million years ago. The lunar crust can be considered to have existed as a single unit since its appearance shortly after the formation of the Moon. The Moon's small size, its lack of a dynamic interior and the relatively great thickness of its crust ruled out all possibility of plate tectonics. Most of the lunar mountains have been produced by impact processes or crustal movement following major impacts. Although some upland areas may be true volcanoes, none of the Moon's mountains have been caused by tectonic-driven crustal compression.

Ninety-eight percent of the Moon's far side consists of highly cratered, mountainous upland. Many of the bigger, generally unflooded far-side impact basins display central massifs and a system of concentric mountain rings. Mare Orientale, a majestic multiring impact basin that lies largely on the Moon's far side, just past the southwestern limb, shows an impressive system of mountains. The Montes Rook mark the Orientale basin's 620-km-diameter inner ring. Its outer ring, 930 km in diameter, is made up of the Montes Cordillera. There is a sharp transition between the smooth plains bounded by the Cordilleras and the radially grooved, mountainous landscape beyond. This impressively sculpted terrain, formed by the erosion and deposition of piles of debris ejected during the Orientale impact, superficially resembles a terrestrial mountainous landscape that has experienced erosion by an ice sheet.

Lava flooding within the near-side impact basins has obliterated many of their internal mountain structures, but here and there peaks do rise abruptly above the flat marial plains, an example is like the nunataks, which protrude through the glaciers of Greenland. The Moon's near side has 18 named mountain ranges and 30 individually named mountains, but there are many more small, unnamed mountain peaks and peak clusters within the maria. Meteoroid erosion has affected the lunar mountains so that their summits are gently rounded; there are no jagged Matterhorn-type peaks in any of the Moon's mountain ranges, contrary to the illusion of spirelike peaks caused by the long shadows they cast when illuminated by a low morning or evening Sun.

Lunar Ice

Some deep craters in the vicinity of the lunar poles never receive direct sunlight; these permanently shadow-filled craters may be considered "polar pits of perpetual darkness." Libration regularly tilts the Moon's poles slightly towards the Earth, so that some of these craters may be observed on or near the Moon's limb, though they are very foreshortened. It is thought that, within some of these perpetually shadow-filled craters, vast quantities of nearly pure water ice lie buried beneath the top layers of the lunar soil on the crater floors. Observations from lunar orbit using neutron spectroscopy (which detects certain types of neutrons emitted by hydrogen atoms) have revealed that the water ice signature is stronger at the Moon's north pole than at the south pole (a signature of 4.6% compared with 3%). The ice is probably concentrated in numerous small areas that cover a total of around 1,850 sq km at each pole. The estimated total mass of the lunar ice is more than 6 billion tons. The ice is thought to have been brought to the Moon by the impact of cometary nuclei, which are very large "dirty snowballs," clumps of water ice (in addition to other volatiles) mixed with rock and dust. The cometary nuclei impacting on the Moon's surface would have been instantly vaporized, but the volatiles would have quickly solidified into a vast cloud of snowflakes if the impact had taken place on the cold, unilluminated hemisphere. As soon as sunlight warmed the surface, the ice lying on the surface would have sublimated again and escaped back into space, but the ice that had fallen in permanently shadow-filled craters would have remained. The water ice at the Moon's poles will become an invaluable resource when human exploration of the Moon recommences later in this century – for human consumption, for industrial purposes and perhaps to produce fuel.

Some of the highest peaks and the rims of many prominent craters at the Moon's north and south polar regions are continually illuminated by the Sun, and have been referred to as "mountains of eternal light." These regions will also attain tremendous importance when humans revisit the Moon in the decades to come, for they will be perfect sites for locating solar arrays, since they receive a continual, uninterrupted supply of sunlight – ideal means to supply the electricity for lunar colonies and industry.

Changes on the Moon

A glance through a telescope eyepiece ought to be sufficient to convince anyone that the Moon is a world whose surface is far less dynamic than the Earth's. The Moon's rugged features are set rigidly in gray rock, and it doesn't appear to change much at all, except for the angle at which features are illuminated by sunlight during the 2-week-long lunar day. The Moon possesses no appreciable atmosphere. No clouds ever drift across the dark lunar skies, no rainfall ever dampens the gray lunar soil and not a breath of wind whips up the dry lunar dust.

Transient Lunar Phenomena (TLP)

Although the Moon appears to be a barren and entirely dead world, there have been occasional reports – mainly from amateur lunar observers – of small-scale transient changes on the Moon's surface called TLP (transient lunar phenomena). TLP have been reported in three basic forms: brief isolated flashes or pulses of light, red- or blue-colored glows and obscurations or darkenings of portions of the lunar surface. Some TLP have been reported to temporarily masquerade as distinct 3-dimensional topographic features. Such bouts of anomalous short-lived activity seem to occur sporadically in a number of specific small areas of the lunar surface. Among the areas most reportedly prone to TLP are the region in and around the bright crater Aristarchus in Oceanus Procellarum, the large flat-floored crater Plato in the lunar Alps, the borders of Mare Crisium, Mare Serenitatis and Mare Imbrium, and a number of craters with faulted floors, notably Alphonsus, Gassendi and Posidonius.

Lunar scientists have been reluctant to accept that our satellite occasionally displays such very obvious signs of activity, and this lack of widespread recognition of the reality of TLP has a great deal to do with the fact that they have been observed mainly by amateur astronomers with limited equipment. Few amateurs have been able to secure a photographic record of TLP, let alone employ anything as sophisticated as a spectroscope during these events when they have occurred. The majority of TLP have therefore been inadequately recorded, at least for the purposes of later in-depth scientific analysis.

Possible Causes of TLP

Features on the Moon appear to change considerably during the 2-week-long lunar day as their angle of illumination by the Sun slowly changes. A topographic feature that may appear bold and conspicuous when illuminated by a low morning or evening Sun can fade into utter obscurity at lunar midday. Conversely, some albedo features are quite invisible when near the morning or evening terminator, but stand out boldly under a high Sun. In addition, the observed brightness of albedo features is dependent on the height above the lunar surface of both the Sun and the observer on the Earth (the selenographic longitude and latitude of the Sun and the observer). A feature observed under a midday Sun on one lunation may appear somewhat brighter (or dimmer) during the following lunation at midday. Some of the Moon's brighter albedo features are noticeable even when they are not directly illuminated by the Sun, but instead bathed in the glow of earthshine. The appearance of Aristarchus, one of the most brilliant lunar features, on the unilluminated side of the Moon, can be striking, and it has been the cause of many reported TLP. The observed brightness of features on the unilluminated hemisphere of the Moon is due almost entirely to reflected earthshine, but the observer can have the distinct (but erroneous) impression that the Moon's surface is glowing of its own accord. Thermoluminescence is caused when certain crystalline materials emit light when they are heated (as excited electrons fall back to their ground state), but the amount of light given off by lunar thermoluminescence is entirely incapable of being detected by amateur instruments, and the process certainly would not explain any reported TLP. There have been reports of TLP (in the form of localized brightenings) taking place at the same time as the arrival of high-energy solar particles emanating from solar flares. These brightenings have been ascribed to the lunar rocks fluorescing as the solar particles impacted on them. Again, the fluorescence of lunar materials is thought to take place, but it is incapable of producing sufficient light to be detected from the Earth.

Finally, the Moon is not painted entirely in shades of gray. Some areas of the Moon do possess delicate, but noticeable, coloration, and the observed intensity of these colors varies with their illumination. A full Moon high in the sky creates the opportunity to scan the Moon and view its large-scale coloration, and a low- to medium-power eyepiece (without a filter) on a telescope that is relatively free from chromatic aberration (a Newtonian reflector, apochromatic refractor or Maksutov catadioptric) is recommended. Achromatic refractors introduce a degree of false color, not usually a problem when the observer is studying topography alone, but distracting if a TLP search is being attempted. One of the most strikingly colored lunar landscapes is a large, quadrilateral area with a reddish hue (unofficially known as "Wood's Spot") that covers around 1,000 sq km to the northwest of the Aristarchus region. Although the lunar southern highlands are bright and generally color-free, there is a surprising amount of color to be made out in the lunar maria. One of the best color contrasts can be seen by viewing the adjacent seas of the reddish Mare Serenitatis and the distinctly blue Mare Tranquillitatis. Even using a good telescope under ideal conditions, the inexperienced lunar observer may not expect to see obvious hues of purple, brown, red, green and blue on the Moon, and their presence may wrongly be ascribed to some kind of anomalous event.

Thus, regularly occurring changes in the angle of illumination and the apparent brightness and colors of features can lead an inexperienced observer to draw the wrong conclusion – that a real change has taken place on the Moon.

Meteoroid Impacts

High temperatures can only be formed at the lunar surface when fast-moving objects impact on the Moon, converting much of their kinetic energy to heat (and light). Some of the short-lived flashes that have been observed in random areas around the Moon may have been caused by meteoroid impacts on the lunar surface. Even a relatively small meteoroid impact would be sufficient to produce a brief flash visible in amateur telescopes, and such an event would be more readily visible were it to take place in the unilluminated portion of the Moon. Flashes of light on the dark hemisphere of the Moon have been visually observed and record-ed by video cameras, and many of these events appear to have taken place during the periods of terrestrial meteor showers – November's Leonids, for example. Nobody doubts that the Moon is hit from time to time by large meteoroids, and that impact flashes bright enough to be telescopically observed are produced, but the suggestion that meteor streams contain individual meteoroidal masses large enough to produce observable impact flashes on the Moon is controversial. It is widely accepted that meteor streams are almost exclusively composed of fragile grains of dust blown out by cometary nuclei, and contain nothing of a size large enough to produce any observable effects on the Moon, however brief. Meteorites found on the Earth have been shown to originate from asteroids, with a few that are known to have been blasted from the Moon and Mars. There are only a few examples of terrestrial meteorite falls having taken place during the peaks of meteor showers, but these have been coincidental events: an examination of these meteorites shows them to be entirely unconnected with the parent body of the meteor stream.

Material thrown up by any sizable meteoroid impact on the Moon would form an expanding shell of material that may itself remain visible for several minutes, especially if the focus of impact happens to lie just beyond the terminator and the ejecta cloud climbs high enough above the lunar surface to be directly illuminated by the Sun. However, no reported TLP flash site has ever yielded a new crater that has been detected from the Earth. Many of the brief flashes may actually have been caused by meteors in the Earth's own atmosphere that have approached along the observer's line of sight.

Internal Mechanisms

Volcanic activity can be ruled out as a cause for TLP. The Moon's crust is cold, solid and immensely thick, and the Moon's surface has not been subjected to true volcanic activity for a couple of billion years. It is highly unlikely that TLP are the product of conventional volcanic activity, since temperatures far in excess of 1,000°C would be required to heat the local rock sufficiently (even melting it) so that it glows visually. It has been calculated that 1 sq km of freshly erupted lava on the Moon (at 1,000°C) would appear as an orange starlike point of magnitude 5.5.

During a couple of hours at sunrise, the lunar surface heats up from a chilly –170°C to a blisteringly hot 130°C (and vice versa at lunar sunset, a fortnight later) – a range of 300°C. Such a sudden temperature change, as the Moon warms up in the morning sunshine and cools down at lunar dusk, takes place twice every lunation. It has been suggested that the thermal shock due to the lunar rocks' expanding as they heat up (and contracting as they cool down) may be a cause of TLP. Although most TLP are reported within a few days of local sunrise or sunset, this may say more about the observers' preference for keeping an eye on the more aesthetically pleasing near-terminator areas than the true occurrence of TLP. Moreover, the Moon's regolith remains at a constant temperature of −35°C beneath a depth of a meter. The processes of thermal shock, restricted to the materials in the very top layers of the regolith, cannot possibly give rise to any effects observable from the Earth.

It is probably no coincidence that most TLP activity takes place in areas that have succumbed to substantial crustal stress and display extensive, deep-seated faulting – around mare borders and on the fractured floors of some craters. Crustal stresses are greater around times of perigee, when the Moon is being pulled more by the Earth's gravity than at other points of its orbit; it has been found that moonquakes are more prevalent at these times. In fact, the tidal pull of the Earth on the Moon is 32.5 times greater than that produced by the Moon on the Earth. Any gases that happen to be trapped in pockets deep inside the Moon's crust may be triggered to escape through faults and fissures into the vacuum of near-Moon space. An escape of gases in itself may not give rise to any immediately observable effects, although it would be detectable spectroscopically. An energetic venting of gas may cause large volumes of the smaller particles in the loose upper few centimeters of the lunar soil to waft high into the lunar skies, causing a temporary obscuration of localized surface features as viewed from the Earth. A large cloud of lunar dust might take on the appearance of a temporary hill for a few minutes, able to block enough sunlight for it to cast a shadow before the material gradually settles back on the lunar surface. Were such an event to take place on the night side of the Moon, just past the morning or evening terminator, the cloud might rise high enough to catch the rays of the Sun, creating a temporary bright spot in the darkness.

Volcanic eruptions on Earth that throw up vast clouds of fine debris are often illuminated with flashes of lightning. Within the dust cloud, a triboelectric charge is built up as the particles rub against each other, and lightning occurs when there are triboelectric discharges. Triboelectric lightning flashes only occur in high-pressure environments (such as the Earth's atmosphere); any triboelectric lunar glows would be far too feeble to be discerned from the Earth. Two other mechanisms might cause a gaseous lunar eruption to glow. When rock crystals are subjected to mechanical strain, an electric potential appears across the crystal faces, causing a glow when this electricity is discharged into surrounding gases. Alternatively, protons and ultraviolet radiation in the solar wind are capable of exciting the atoms in the vented gases, ionizing them to produce a glow discharge.

Physical Lunar Changes

In addition to elusive bouts of short-lived TLP activity, lunar changes of a permanent nature have also been suggested; there are a number of alleged cases. A great deal of controversy has surrounded claims of permanent change visible from the Earth, and evidence in their favor has often been weak and unreliable. Advocates of lunar change, attempting to establish a degree of respectability in their field, have been unfortunately hindered by the opinions of cranks and the spurious observations of charlatans. Photography provides the firmest kind of evidence, yet there has not been a single reliable photograph taken from the Earth that proves that permanent lunar change has taken place.

Crater Linné, 100 km off the western coastline of Mare Serenitatis, is a tiny 2.4-km-diameter pit surrounded by a bright circular ejecta collar around 10 km in diameter. Linné is easily visible in binoculars and small telescopes. When the area is illuminated by a low early morning or late afternoon Sun, a 150mm telescope should be just sufficient to resolve the tiny crater itself. It has been claimed that Linné was once a much larger crater – as large as 10 km across – and somehow it had shrunk in size. These assertions were sensationalist and based on poor science. A close look at the body of evidence pointing towards a once-large Linné will show that the case rests on flimsy foundations. Old general maps, full of other glaring errors in cartography, can hardly be used as evidence, and many historical observations of Linné are contradictory. Linné has not changed appreciably since its formation, perhaps several hundred million years ago.

Around 150 km off the western shore of Mare Fecunditatis lie a pair of small craters positioned side-by-side on the marial plain. The smaller of the two is Messier, a crater elongated in an east–west direction, measuring 9×11 km. Its neighbor, Messier A, measures 13×11 km, and has a more distorted shape owing to an extended (nearly double) western rim. A straight and narrow double ray extends like a searchlight beam from Messier A across to the hills marking the mare border. The Messier pair, fraternal rather than identical twins, make a pleasant sight through any telescope. They provide a good example of how the appearance of a lunar feature can alter considerably during a lunar day because of the changing angle of illumination, as anyone who has observed them under a variety of lighting conditions can testify. Throughout the lunar day, the Messier twins can appear to shrink in diameter, assume distinctly nonelliptical shapes, alternate in size (Messier sometimes looking as though it is the larger) or even merge into one; the rays that emanate from Messier A also undergo changes in apparent brilliance. These are all tricks of the light that can lead the inexperienced observer to conclude that a real metamorphosis has taken place.

Physical change does happen on the Moon, but on a scale that is usually unobservable through telescopes on Earth. Hard photographic evidence from cameras in lunar orbit shows that things aren't altogether static on the Moon's surface. For example, close-up orbiter photographs of the central hills of the crater Vitello show two sizable boulders, the larger of which is around 25 m in diameter; both objects made distinct tracks in the lunar soil as they rolled down the slope. The boulders must have been jarred from their original perches by a large moonquake – a vigorous shaking of the lunar crust that perhaps resulted from a large meteoritic impact close to Vitello. Using photographs alone, it is impossible to determine the precise date on which the boulders made their short journeys: they could have rolled a moment before the probe took its pictures, or they may have tumbled down the hill many thousands of years ago. The incredibly slow pace of erosion on the Moon, chiefly caused by a steady drizzle of micrometeorites, means that the tracks impressed in the soil are likely to remain fresh-looking for thousands of years to come.

High-resolution photographs taken by orbiting probes have uncovered many more examples of small, localized physical changes due to migrating surface debris. Natural lunar soil creep (gradual downhill migration) happens because moonquakes and occasional large meteoroid impacts make the Moon's solid crust shudder, causing the overlying regolith to settle into a more comfortable position.

For example, the base of Hadley Rill, a large lunar canyon on the eastern border of Mare Imbrium, is covered with a thick blanket of soil embedded with sizable boulders that have fallen from the higher slopes.

Another mechanism of dust transport may be taking place on the Moon. Observations of diffuse glows on the lunar horizon near the terminator, made from lunar orbit, plus amateur observations of the "graying" of normally intensely black shadow near the Moon's terminator, may indicate the presence of transient dust clouds suspended above the lunar surface. The lunar soil has low conductivity, is readily charged and retains its electrical charge for long periods. It is possible that charged particles in the solar wind, along with ultraviolet radiation, electrostatically charges the lunar soil, resulting in the levitation (perhaps to heights of several kilometers) and the transportation of small dust particles along the slowly moving boundary of the terminator. With its extreme clinginess, this charged material may present a hazard to future human lunar exploration.

Chapter 2

The Measure of the Moon

Measured across the equator, the Moon has a diameter of 3,476 km, roughly equivalent to the breadth of the continental United States. Averaging out the global topography of the lowland basins and mountainous highlands, the figure of the

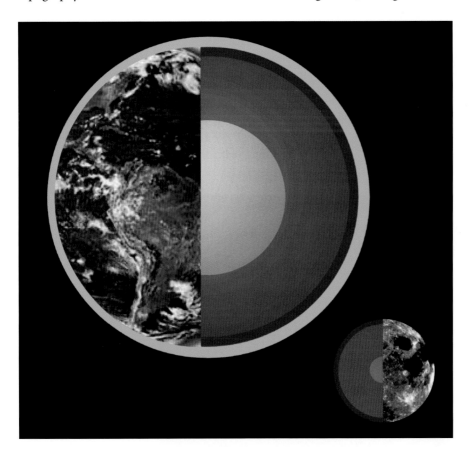

Fig. 2.1. Comparison between the Earth and Moon, showing their internal structures. The Earth has a relatively thin, mobile crust underlain by a hot, dynamic mantle and a large hot, iron, core. The Moon has a relatively thick, static crust and a warm interior, with a very small core. Credit: Peter Grego

Moon is not a perfect sphere – it is a geoid, a sphere flattened on its polar axis by about 2.2 km, with a slight equatorial bulge of a few hundred meters pointing in the direction of the Earth.

With a volume of just under 22 billion cubic km, the Moon has just 2% of the Earth's volume. However, the Moon has a mass of 73,500 trillion tons, which amounts to just 1.2% the mass of the Earth. Therefore, the Moon is 40% less dense than the Earth. If both the Earth and Moon were shrunk to the size of tennis balls, the wet blue one would weigh 550 g and the dusty gray one would weigh 330 g. The strength of the Moon's gravity at its surface is 16.5% of the Earth's. On the Moon, an object weighs only about one sixth its terrestrial weight, and when dropped will fall to the lunar surface with 6 times less acceleration due to gravity. To leave the surface of the Moon and enter into lunar orbit (or beyond), an object must travel at an escape velocity of 2.38 km/second. Escape velocity can be reached through the gradual acceleration produced by a rocket engine, or by sudden means – say, during an asteroidal impact that launches some of its ejecta into space.

The Moon's Orbit

Planets move around the Sun in elliptical orbits, and their satellites follow elliptical orbits around them. An ellipse is a closed curve with two focal points lying on its main axis; the Earth lies at one focal point of the Moon's orbital ellipse. Drawn to scale, the Moon's orbit around the Earth appears almost circular, with the Earth positioned very close to the center of the circle. Careful measurement will reveal that the figure is really an ellipse with a mean eccentricity of 0.055 (the ratio of the difference between the major and minor axes to the major axis), with the Earth lying slightly to one side of the center, positioned over a focus of that ellipse.

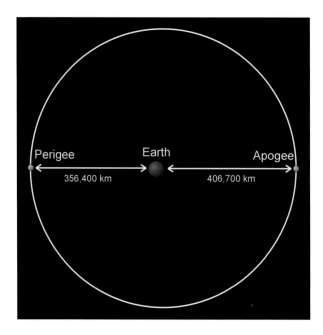

Fig. 2.2. Orbit of Moon.

Measured with respect to the stars, the Moon takes 27d 7h 43m to make one revolution around the Earth, a period called the sidereal orbital period.

It is not absolutely correct to claim that the Moon revolves around the Earth. In fact, both the Earth and Moon revolve about their common center of gravity, a point called the barycenter. If the Earth and Moon were equal in all respects, then the barycenter would be positioned in space exactly between the Earth and Moon. But the Moon's mass is only 2% that of the Earth's, which offsets the barycenter considerably in the Earth's direction, so much so that the common center of gravity is actually located within the Earth's mantle, around 4,700 km from the center of the Earth. Plotted to scale and viewed from above, the path of the Moon around the Earth is always concave to the Sun.

Orbiting at an average distance from the Earth of some 384,401 km, the Moon lies about 30 times the Earth's diameter away. Light (and other forms of electromagnetic radiation) takes an average of 1.3 seconds to cross the gulf of space between the Moon and the Earth. Radar signals bounced off the Moon enable its distance to be determined to an accuracy of under half a kilometer, and the Moon's libration (the slow rocking on its own axis) has also been detected. The Moon's surface does not reflect radio waves as strongly as it would were it covered with large expanses of solid rock, and it was known to be covered in a thick layer of soil long before the first soft-landing probes touched down on its surface. The most accurate means of measuring the distance of the Moon is to aim short pulses of laser light at a reflecting point at a known location on the lunar surface and accurately time the returning light signals. Measurements made by aiming lasers at the passive laser retro-reflectors left at the Apollo landing sites give a Moon distance accurate to within a few meters, and observations over the years have demonstrated that the Moon's mean distance from the Earth is slowly increasing.

At its furthest from the Earth, at apogee (apo: far; ge: Earth), the Moon can be as distant as 406,700 km; when closest to us at perigee (peri: near) it can approach

Fig. 2.3. The difference in apparent size of the Moon near perigee and apogee. Credit: Mike Goodall

356,400 km. The average angular diameter of the Moon as seen from the Earth's surface is 31′ 05″. At perigee its angular diameter is 33′ 29″, and at apogee it measures 29′ 23″. Having an apparent diameter about 12% larger at perigee, the apparent area of the perigee Moon is a huge 29% larger than at apogee. The imaginary line that joins the points of apogee and perigee – effectively the major axis of the Moon's elliptical orbit – is called the "line of apsides". The line of apsides rotates (with respect to the stars) in a prograde fashion every 8.85 years.

The plane of the Moon's orbit around the Earth is inclined to the plane of the Earth's orbit around the Sun (the plane of the ecliptic) by 5° 8′ 43″. The two points at which these planes intersect are called the ascending and descending nodes. The ascending node is the point on the ecliptic where the Moon moves to the north of the ecliptic; the descending node marks the point where the Moon moves to the south of the ecliptic. The line of nodes rotates (with respect to the stars) in a retrograde fashion every 18.61 years. In an astronomical sense, the recession of the line of nodes is important, since solar and lunar eclipses are dependent on the nodes' position in relation to the Sun and the Earth.

The inclination of the Moon's orbit to the ecliptic, combined with the recession of the line of nodes, means that every 18.61 years the Moon attains its most southerly declination below the ecliptic, in the constellation of Sagittarius. Half an orbit later, the Moon attains its most northerly declination, on the border of Gemini and Taurus. When the full Moon is positioned near the extreme southerly point of its orbit, it barely clears the southern horizon at its culmination around midnight in June, when viewed from northern temperate latitudes. Viewed from New York (41°N), an extreme southern Moon reaches an altitude of 20° above the southern horizon. From London (52°N) it nudges to an altitude of 9°, just 18 times its own apparent diameter. From Lerwick in the Shetlands (60°N), the extreme southerly full Moon barely lifts above the horizon for a couple of hours as it appears to roll across the distant southern horizon. A similar situation, with exactly opposite circumstances, happens in the southern hemisphere. From southern temperate latitudes, the full Moon appears very low above the northern horizon during December. From Alice Springs, Australia (24°S), the most northerly midsummer full Moon rises to an altitude of 38° above the northern horizon. From Melbourne (37.5°S), the Moon culminates at an altitude of 24.5°. However, the extreme northerly full Moon doesn't appear to roll along the horizon as it does from the northern latitudes of Lerwick, since there are no mainland locations near the 60°S line of latitude. The most northerly and southerly terrestrial latitudes at which the Moon can appear directly overhead are about 28°N and 28°S, respectively. From the latitude of London, the Moon can never appear closer than around 23° to the zenith. From the latitude of New York, this distance is reduced to around 12°.

Gravity and Tides

The gravitational pull of the Moon is the main cause of the Earth's ocean tides. Two tidal bulges in the Earth's oceans are produced, and a line projected through the peaks of these bulges points in the general direction of the Moon. The tidal bulge closest to the Moon is drawn towards the Moon because the strength of the Moon's gravitational attraction is greatest at the sublunar point; this is known as the direct tide. The solid globe of the Earth is also pulled upon by the Moon, but to a lesser degree, and of course the shape of the Earth does not deform appreciably. The Moon's gravity is at its weakest at its furthest point from the Moon, on the opposite side of the Earth, and centrifugal forces produced by the rotation of the Moon and Earth around the system's barycenter give rise to an opposite tidal bulge.

As the Earth rotates underneath the two tidal bulges, an alternating rise and fall in the level of the sea with respect to the land is produced. Most of the Earth's coastlines experience a tidal cycle of two high tides (one direct tide and one opposite tide) and two low tides (the troughs between the high tides) per day. Regional geography determines the magnitude of the tides experienced on different coastlines: some places have very unequal high tides, and others experience just one high tide per day. If the Moon were stationary in the sky, the tidal cycle would last precisely 24 hours, but because the Moon moves in its orbit, culminating in the sky on average 50 minutes later each day, the tidal cycle is about 24h 50m long.

Earth's oceans are also pulled upon by the gravitational attraction of the Sun, and the tides caused by the Sun amount to about half the height of the Moon's tidal bulges. When the Moon is near new or full phase, the Earth, Moon and Sun are aligned – a phenomenon known as syzygy. At syzygy, the Sun's gravitational attraction complements that of the Moon, producing the most pronounced tidal bulges. Called spring tides, these highest and lowest tides of the month occur every two weeks, around new and full Moon. When the gravitational pull of the Moon and Sun are at right angles to each other, the tides work against each other to produce the least variation in high and low tides. These are called neap tides, and they occur during the first- and last-quarter lunar phases.

Considerable angular momentum – a product of the masses and velocities within a system – is possessed by the Earth and Moon. The law of conservation of angular momentum states that in a system like that of the Earth and Moon, the total momentum remains constant. Friction between the ocean tides and land masses causes the Earth to lose angular momentum, and, as its rotation slows, the length of the terrestrial day gradually increases. The rate of change in the Earth's secular rotation actually amounts to just 2.3 milliseconds per century (23 seconds every million years), large enough to be measured against atomic clocks. Another consequence of the friction between the Earth's crust and the oceans is the displacement of the tidal bulge slightly forward of the direction of the Moon. The additional gravitational attraction of this displaced tidal bulge actually tugs on the Moon and imparts angular momentum to it. As the Moon's angular velocity increases, so does its distance from the Earth. The Moon is receding from the Earth by about 3.8 cm per year (38km every million years), a figure verified by precisely measuring the time taken for laser beams to reflect from instruments placed on the Moon's surface.

Secular and Periodic Perturbations

The ellipse of the Moon's orbit is not static and frozen rigidly into the fabric of space. It experiences distortions, principally caused by the gravitational pull of the Sun, and to some degree by the shape of the Earth (an even more flattened sphere than the Moon) and the gravitational attraction of the other planets. Perturbations by the Sun's gravity compel the Moon's orbital ellipse to rotate slowly in space. The line of apsides (the ellipse's major axis) advances eastwards against the stars by around 3.3° each month; this means that the period from one perigee to the next is longer than the Moon's sidereal period by more than 5.5 hours. One complete (prograde) revolution of the line of apsides takes 8.85 years. In addition to the slow advance of the line of apsides, the other major secular perturbation is seen in the recession of the line of nodes (the line joining the ascending and descending nodes, as described above), which moves westwards along the ecliptic by 1.6° each month. The line of nodes takes 18.61 years to make one complete retrograde revolution around the ecliptic. The recession of the line of nodes has important consequences for the timing of solar and lunar eclipses.

When the major axis of the Moon's orbital ellipse points towards the Sun – a situation that occurs every 4.4 years when the Moon is either full or new and is around apogee or perigee – the Sun's attraction will stretch out the ellipse, making it slightly more eccentric. Over 2 years later, the ellipse's major axis has rotated so that it is perpendicular to the Sun's direction; apogee or perigee then coincides with either the first or last quarter, and by this time the orbital ellipse will have been gradually pulled back into a less eccentric shape. This periodic gravitational tugging is a phenomenon known as evection, and causes the eccentricity of the Moon's orbit to vary between 0.0432 and 0.0666, a complete cycle recurring every 87.1 years. As a consequence of evection, the Moon can be displaced in longitude by as much as two and a half times its own diameter, a displacement so great that it was observed in ancient times and recognized by Hipparchus.

Variation is the second most important periodic perturbation of the Moon's orbit. It causes a periodic displacement in the Moon's apparent longitude in the sky, amounting to as much as 39.5' (a little more than the Moon's apparent angular diameter), and results from the interplay of the gravitational attraction of the Earth and Sun. At times when the attraction of the Earth and Sun is combined (at new or full Moon), the effect of variation cannot be detected. Variation is not evident the first or last quarter phase because the gravitational forces of Earth and Sun are then perpendicular, canceling each other out. It is only when the Moon is moving towards or away from either the new or full Moon that displacement due to variation can be observed, as the Moon's motion is being either helped or hindered by the combined attractions of the Earth and Sun.

Libration

Locked in a synchronous rotation, the Moon keeps the same face turned in the Earth's direction as it orbits. A phenomenon known as libration – a term describing a rocking motion of the Moon's globe that is too slow to be observable in real time at an eyepiece – means that from the Earth a total of 59% of the Moon can be observed over a period of time. The remaining 41% of the lunar surface constitutes the side that is permanently hidden from the terrestrial observer.

Libration has two fundamental modes: optical libration and physical. Optical libration results from the ever-changing aspect (within strict limits) of the Moon's presentation to the terrestrial observer, combined with the position from which the Moon is viewed. Physical libration is an actual wobble of the Moon about its own center of gravity, and it is a real, though exceedingly small, oscillation, rather than an apparent one.

Optical Libration

Optical libration causes the most obvious effects on the apparent position of features in and around the edge of the Moon, and there are three kinds: libration in longitude, libration in latitude and diurnal libration.

Libration in longitude is produced by the ellipticity of the Moon's orbit, which leads to a variation in its angular velocity. The Moon accelerates on its way towards perigee and then slows down on its outward leg to apogee, but at all times it maintains a more-or-less constant rate of rotation upon its own axis. This results in the Moon's appearing to rock from side to side during the lunar month. Maximum libration in longitude occurs about a week after perigee and apogee – at perigee and apogee there is no libration in longitude. Take the Moon as it completes a "fast" quarter section of its orbit after perigee: it will cover this quarter orbit in less time than it took to make a quarter of a turn on its own axis. The Moon's axial rotation at this point is lagging behind, and the mean center of the Moon's disc will appear displaced towards the western limb. Far-side features will therefore be brought onto the eastern edge of the Earth-facing hemisphere, and they can be observed if the illumination is right. Half an orbit later, when the Moon is moving away from apogee in a slower section of its orbit, the opposite situation occurs. In this case,

Fig. 2.4. Libration in longitude is caused by the fact that, while the Moon's rotation on its own axis is constant, the Moon's orbit around the Earth is elliptical, with the Earth located at one focus of this ellipse. At perigee (P), there is no libration in longitude. A quarter of its orbit later (X), the Moon has moved in its orbit a little more quickly than at its mean rate, and as a result the mean center of the Moon's disc is displaced to the west, libration favoring the Moon's eastern limb. As the Moon slows towards apogee (A), its axial rotation has caught up with its orbit, and there is no libration in latitude. A quarter of its orbit later (Y), the Moon's axial rotation has outpaced the orbit and the mean center of the lunar disc is displaced to the east, resulting in a good libration for features on the western limb. Credit: Peter Grego

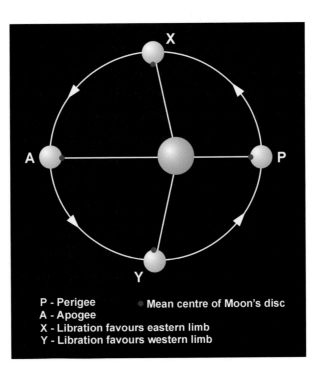

P - Perigee
A - Apogee
X - Libration favours eastern limb
Y - Libration favours western limb

● Mean centre of Moon's disc

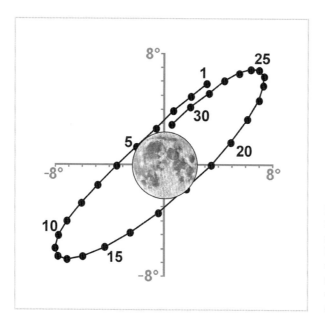

Fig. 2.5. Typical chart of a month's libration (in this case, April 2003) showing how the most favored point of libration at the Moon's edge gradually moves anti-clockwise around the lunar disc. Credit: Peter Grego

the Moon has made a quarter – turn on its axis before it has made a quarter of an orbit around the Earth, and the apparent displacement of the Moon's surface features will be towards the east. Features that are on the mean far side will make their appearance on the Moon's western limb. Twice every anomalistic month (the lunar month from perigee to perigee), libration in longitude creates a maximum east–west displacement on opposite sides of the Moon's edge. The displacement varies between about 4.5° and 8.1° because of gravitational perturbations in the Moon's orbit caused by the Sun.

Matters would be simpler if the Moon's axis of rotation were perpendicular to its orbital plane around the Earth. But the lunar poles are actually tilted 6.4° to its orbital plane, and the axial tilt remains fixed in one direction in space. Each month, this enables the observer to peer a little over the mean northern limb, and 2 weeks later to view beneath the mean southern limb of the Moon. Libration in latitude, as this phenomenon is called, displaces the mean center of the Moon north–south by between 6.5° and 6.9°. The observer's view of the Moon also changes because the Earth itself is revolving, carrying the observer around with it. The extent of this diurnal libration depends on the observer's actual location on the Earth, and is restricted by the Earth's apparent diameter of some 2° as seen from the Moon. Diurnal libration can appear something like the parallax effect produced by an alternate left- and right-eye views of a tennis ball placed 10 m away. The maximum diurnal libration amounts to just under a degree in lunar longitude at moonrise and moonset.

Features that lie within the Moon's libration zones around the edge of the lunar disc on either side of longitudes 90°E and 90°W are prone to the effects of libration. Over time, all the features within the libration zones eventually make an appearance past the lunar limb. Whenever the Moon is observed, a narrow crescent of the Moon's far side (ie., further than the 90°E or 90°W line) is likely to be presented to the observer to some extent, although this region is not always visible because it might be on the limb of the unilluminated hemisphere. The "mean"

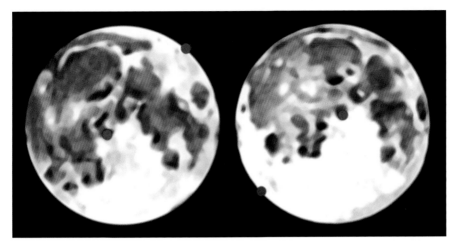

Fig. 2.6. A comparison between two extremes of libration, showing the most favored direction of libration and the offsetting of the center of the Moon's disc. At left, libration favors the northeastern limb, and mare Humboldtianum is very favorably placed. At right there is a favorable southwestern libration, with Mare Orientale on display. Credit: Peter Grego

Moon – a full Moon with the 90°E and 90°W lines of longitude positioned exactly around the Moon's edge, as depicted on most lunar charts and atlases – has rarely been observed.

Optical librations and the Moon's phase determine the practical visibility of the Moon's limb features. For example, Mare Orientale lies largely past the 90°W line near the Moon's southwestern limb. The feature is not at all visible when there is a libration favoring the Moon's eastern limb, and even when libration favors the southwestern limb, a good libration of around 7° when the Moon is past full phase is needed in order to view the whole of the mare as a thin, dark sliver near the edge of the Moon. Near-side, near-limb features are best observed when they are present

Fig. 2.7. The 6-day-old Moon. Libration favors the Moon's northeastern limb, where Mare Humboldtianum can easily be discerned. Credit: Dusko Novakovic

at a favorable libration. A large crater like Gauss, near the northeastern limb, can appear exceedingly foreshortened, and its floor is difficult to observe at an unfavorable libration, but it opens up very nicely at a good libration. There is little point attempting to observe features on the floor of Mare Humboldtianum, which lies near the northeastern limb on the near side, when there is a strong libration favoring the southwestern limb. The apparent shapes of features positioned nearer to the center of the lunar disc are not affected as much by libration, although libration affects the timing of their appearance at the terminator. For example, the crater Ptolemaeus, near the center of the disc on the Moon's central meridian, will appear on the lunar morning terminator at precisely first quarter phase when there is no libration in longitude (around apogee and perigee). However, libration can cause Ptolemaeus to appear on the morning terminator of a distinctly crescent Moon, as well as on a considerably gibbous Moon.

Viewed from the Moon, the Earth appears to hang forever in the same part of the sky, the background stars of the zodiacal constellations appearing to slip behind the Earth once a month. But the Earth isn't completely static, as it slowly moves around in a small ellipse in response to the Moon's librations in longitude and latitude. Observed from an area within the libration zones, the Earth will appear to rise above the horizon and slip beneath it once again every anomalistic month.

Physical Libration

Physical librations of the Moon around its own center of gravity never amount to more than two minutes of arc in lunar coordinates, an angular displacement of under a second of arc at the center of the Moon's disc. Because their magnitude is so tiny, physical librations do not have any appreciable impact upon the observability of lunar features.

Lighting Effects

Shining only by reflected sunlight, the Moon appears silvery and brilliant at night because it is viewed against a dark background. In reality, the surface of the Moon is rather dark. In fact, the Moon is one of the least reflective worlds in the entire Solar System, with an average near-side albedo (a measure of reflectivity) of just 0.07, meaning that only 7 out of every 100 photons that hit the lunar surface manage to bounce back into space. From the Earth, the intensity of the light of the full Moon is only 0.25 lux (one lux is the brightness of a burning candle placed a meter away). Shining at magnitude –12.7, the full Moon is about 400,000 times dimmer than the Sun. Being spherical, the full Moon reflects the greatest proportion of light in the direction of the observer from areas nearer its center. If the Moon were a perfectly smooth sphere, its average brightness would be raised considerably. However, the rough and irregular lunar surface throws up shadows, and this causes the brightness of the half-Moon of first or last quarter phase to be just one-tenth of the full Moon's brightness.

Viewed from the near-side surface of the Moon, the full Earth appears as a bright blue and white globe with an apparent area more than 13 times that of the full Moon. Since the albedo of the Earth averages 0.39 – more than 5 times the brightness of the Moon – the full Earth viewed from the Moon is nearly 70 times brighter than the full Moon seen from the Earth. Viewed from the Moon, the Earth goes through a complete cycle of phases during the same period as the Moon's cycle of phases. The two bodies' mutually observable phases are opposites of one another; for example, when the Earth appears full from the Moon, the Moon appears new from the Earth; when the Moon is a waxing crescent, the Earth is in a waning gibbous phase.

Earthshine

How could a spectroscope that is pointed directly towards the surface of the life-less Moon possibly detect water, molecular oxygen and chlorophyll, the latter indicating the presence of land plants? These substances can actually be detected by analyzing the spectrum of sunlight that has been reflected from the Earth onto the Moon and back to the Earth. The ashen glow of this twice-reflected light is called "earthshine," a faint glow on the unilluminated portion of the Moon, often visible with the unaided eye and striking through binoculars. Earthshine is most obvious when the Moon is a thin waxing or waning crescent in a reasonably dark sky. This spectacle – one of the prettiest sights visible without optical aid – is sometimes called "the old Moon cradled in the young Moon's arms." Viewed from the near side of the Moon at this phase, the Earth would appear as a brilliant, near-full disc. With the unaided eye, earthshine can be discerned for several days until around the first quarter phase and picked-up again in the last quarter of the lunar month, as the Moon wanes into a thin crescent. Earthshine is best seen when the Moon is high in a relatively dark sky, with the Sun positioned far beneath the horizon. The best circumstances for northern hemisphere observers occur in the evening skies of spring and the morning skies of autumn.

The observed intensity of earthshine depends on the phases of the Moon and the Earth and the reflectivity of the Earth. The latter factor is somewhat variable, and depends on the amount of cloud cover and how the bright terrestrial polar regions, continents and oceans are arrayed on the side of the Earth that is reflecting sunlight onto the Moon. Regular and attentive observers of the Moon will notice that the intensity of earthshine sometimes appears to be particularly prominent, even taking into account the phase of the Moon and local conditions. This is not an illusion, as photometric measurements have shown that the brightness of earthshine has a cyclic component that varies with the terrestrial seasons. Earth's reflectivity has been measured at around 20% brighter between October and July.

Phases

As the Moon revolves around the Earth it goes through a complete sequence of phases. The Moon shines only by reflected sunlight, and its phases are produced by the angle made by the Earth, Moon and Sun. The cycle of phases proceeds from new Moon (when the Moon is in line with the Sun and the Earth), and "waxes" through crescent, first quarter (when the Moon makes a right angle between the Earth and Sun) gibbous phase to full Moon (when the Moon is on the opposite side of the Earth from the Sun). After full, the Moon "wanes" through gibbous, last quarter, crescent and back to new Moon once more. The period from one new Moon to the next is called the synodic month, and averages 29d 12h 44m in length – it actually varies between 29d 6h 35m and 29d 19h 55m owing to gravitational perturbations from the Sun. The difference between the length of the Moon's sidereal orbital period and the synodic month comes about because the Earth is in orbit around the Sun, and the Moon must travel a short part of its orbit further after each sidereal month in order to catch up with the Sun's apparent changing position with respect to the background stars.

Observability of the Phases

Moving from west to east in the sky by its own diameter each hour, the Moon traverses about 13° every day. Every 29.5 days it goes through a complete cycle of phases, waxing from new Moon to full Moon and waning to new Moon once more, the synodic month, or lunation. The Moon's path lies close to the ecliptic, the course followed by the Sun through the zodiacal constellations through the year. A first quarter Moon appears in a part of the sky where the Sun will be located 3 months later. When the Moon is full, it lies on the opposite side of the sky from the Sun (i.e., the approximate position where the Sun will be in 6 months' time). Exact alignments of the Sun, Earth and Moon produce lunar eclipses, when the Moon passes through the Earth's shadow, but most of the time the Moon passes above or below the Earth's shadow without being eclipsed. When the Sun is at its lowest point above the horizon at southern culmination, at winter solstice, the full Moon in Taurus or Gemini reaches its heights at around midnight. At summer solstice, the midday Sun reaches its highest point above the southern horizon, and the full Moon appears very low near the southern horizon at midnight. The last-quarter Moon appears in a position where the Sun was located 3 months previously. This has implications for observing the Moon, especially from temperate locations: the higher the Moon is in a dark sky, the better the view through the telescope because of increased contrast and less trouble from atmospheric degradation of the lunar image.

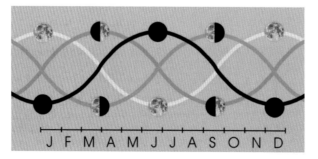

Fig. 2.8. Each month, the Moon follows an orbit close to the plane of the ecliptic, and its height above the horizon varies, just as the Sun varies in altitude through the year. This graph represents the varying altitude of the Moon in its four main phases at southern culmination from northern temperate regions. For example, winter Full Moons are high, while summer Full Moons barely nudge above the southern horizon at midnight. Credit: Peter Grego

Intervals between the times of successive moonrises each night are called retardations, and they average less than an hour, the Moon rising around 50 minutes later each day. If the Moon's orbit were circular and the plane happened to coincide with the celestial equator, all retardations would be around the same length. However, retardations vary in length throughout the lunar month, and for a given phase they vary according to the time of the year at which the Moon is observed. The observer's latitude also plays an important part in how greatly lunar retardation is seen to differ from the norm.

Shadowplay

Every solid object in the Solar System has one side directly illuminated by the Sun, and the other in darkness. The terminator is the line that divides the object's sunlit side from its dark side – literally, the line where sunlight terminates. Of course, when a satellite is immersed in its primary's own shadow and becomes eclipsed, the terminator disappears, since there is no direct sunlight falling onto the eclipsed body. In addition, the dark side of many satellites is often not truly dark; it can be illuminated by sunlight reflected onto it by their primaries (earthshine in the case of the Moon). Some planetary terminators – like that of the gas giant Jupiter or the cloud-swathed Venus, are not clear-cut, and blend gradually from one hemisphere to the other. Since the Moon has no atmosphere, it always displays a sharply defined terminator, and the dividing line between the illuminated and unilluminated lunar hemispheres is unambiguously delineated. The terminator-crossing regions of strong and varied relief are the most clearly defined, as the rims of craters and mountains facing the low Sun reflect the light strongly towards us. Flatter areas that present a smooth shallow angle to the Sun, such as the marial plains, do not show such a stark contrast of light and dark at the terminator. While the terminator in these areas can be traced at low power through the telescope eyepiece, high magnification will reveal a gradual fade from the day to night hemispheres. This effect is most evident in photography, where, to prevent burning out the brighter areas on the Moon's disc in sunlight, the terminator may be underexposed. Since the Sun is an extended source of illumination, half a degree wide, the fade at the terminator is more pronounced than if the Sun were a single point source. At the Moon's equator, the Sun takes about an hour to set, and this produces a gradient of illumination at the terminator. At the Moon's equator, the terminator proper – the place where a rising or setting Sun can be seen – is about 15 km wide (half a degree in longitude), and the Sun takes an hour from first contact with the horizon to its setting. It follows that at the equator the speed of the Moon's terminator is, on average, about 15 km per hour (a brisk jogging speed) taking 29.5 days to make a complete circuit about the Moon, with its circumference of 10,920 km, from one sunrise/set to the next.

Most relief detail can be seen along the lunar terminator because the low angle of illumination throws up shadows from all topographic features. Features just a few tens of meters high can be observed because of the shadows they cast. Countless hours can be spent simply looking at the lunar landscape along the terminator, from the myriad spectacular crater fields of the lunar southern uplands to the vast marial lava plains in the north, with their more subtle detail like domes, wrinkle ridges and rilles.

Retardations

In equatorial regions of the Earth, in which the angle at which the ecliptic (and the Moon's orbital plane lying nearby) intersects the eastern horizon varies the least, the interval between successive moonrises ranges from around 30 minutes to an hour. For observers in temperate latitudes, the range of lunar retardation is far greater, from a few minutes up to as much as an hour and a half. This is because the angle at which the ecliptic intersects the local horizon from temperate latitudes varies greatly. Around the days of full Moon in autumn the ecliptic is at a very shallow angle to the eastern horizon, causing the Moon to rise just 15–20 minutes later each evening. The Harvest Moon is the full Moon nearest the date of autumnal equinox on 23 September, and was so named because farmers were able go about their agricultural business from dusk to dawn assisted by a bright Moon that made its appearance only slightly later each evening. In spring, the ecliptic is at a steep angle to the eastern horizon, meaning that around the time of full Moon, its period of retardation is greatest, and consecutive evening rising times are most widely separated.

Atmospheric Phenomena

Lunar rainbows are created under the same circumstances as rainbows produced by the Sun: a colored arc of 42° radius situated directly opposite the light source. Lunar rainbows have a maximum brilliance of a mere 1/400,000 of their daytime counterparts, and they are seen less frequently because they are dim and rather hypochromic. A bright, pearly white region, the lunar corona, can often be seen around the Moon. Unlike the solar corona, which is part of the Sun's atmosphere, the lunar corona is simply caused by the reflection and diffraction of moonlight amid water droplets in the Earth's lower clouds. One or two (rarely, three) rainbow-colored halos are sometimes observed to encircle the Moon; this is caused by the diffraction of moonlight by ice crystals in the Earth's upper atmosphere. Sometimes the lunar halo may be host to parselene, diffuse images of the Moon located 22° from either side of it, caused by moonlight's refraction among ice crystals in Earth's upper atmosphere.

Eclipses

All solid objects in the Solar System illuminated by the Sun cast a shadow into space. Since the Sun is an extended light source, some half a degree in diameter as seen from the Moon, the shadow is made up of two components – the umbra and the penumbra. The umbra is the darkest part of the shadow and is a cone of deep shadow formed by the external common tangents of the Sun and the object. The penumbra, a shadow cone whose diameter increases with distance from the object casting the shadow, surrounds the umbra and is formed by the internal common tangents of the Sun and the object. Viewed from inside the penumbral shadow, only part of the Sun would appear covered by the object. The intensity of the darkness of the penumbra gradually rises from near zero at the penumbra's outer edge to almost total shadow near the outer edge of the umbra.

An umbral shadow cone over 1.3-million-km long is cast by the Earth. At the distance of the Moon; a section through the Earth's umbra is over 9,000 km across, and the penumbra is around 17,000 km across. The Moon itself casts an umbral shadow that occasionally – and then just barely – exceeds the Earth–Moon distance.

From time to time the Moon appears to pass in front of the Sun, producing a solar eclipse. Sometimes the Moon enters the Earth's shadow to become eclipsed itself. Since the plane of the Moon's orbit is inclined around 5° to the ecliptic, exact alignments of the Sun, Earth and Moon do not occur every month; this would only happen if the Moon's orbit lay along the ecliptic plane. Orbital geometry dictates that the Earth can experience a maximum of either five solar and two lunar eclipses, or four solar and three lunar eclipses, within any single year.

Solar Eclipses

Solar eclipses can only occur when the center of the Sun is less than 18.5° from one of the lunar nodes (the point where the Moon's orbital plane intersects the ecliptic) at New Moon. One of the Solar System's greatest coincidences is the fact that both the Sun and Moon have an apparent angular diameter of around half a degree as viewed from the Earth's surface. The Moon's apparent angular diameter varies

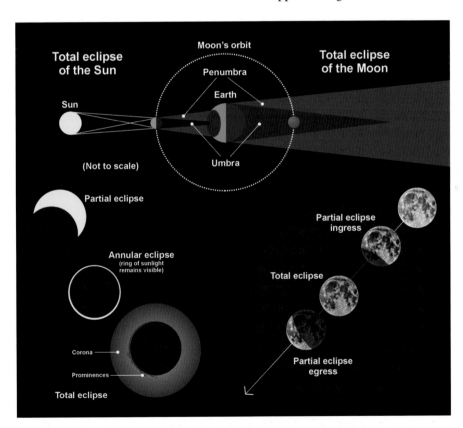

Fig. 2.9. Illustration of the dynamics of solar and lunar eclipses and their phases. Credit: Peter Grego

between 33′ 29″ at perigee and 29′ 23″ at apogee, and the Sun's apparent angular diameter varies between 32′ 36″ at perihelion and 31′ 32″ at aphelion. The fact that the apparent angular diameter of the Moon can be slightly smaller than that of the Sun means that not all solar eclipses include a period of totality. On those occasions when the Moon is observed to pass directly in front of the Sun, yet is too small to produce a total eclipse, an annular eclipse takes place. At midevent the Moon appears as a black circle surrounded by a brilliant ring of sunlight.

A total eclipse of the Sun is one of the most spectacular sights that nature has to offer. For a few moments the disc of the Sun is completely hidden by the Moon. The observing site is engulfed in darkness as temperatures drop, and a momentary silence is soon broken by gasps of astonishment from its viewers. During totality, the brighter stars and planets become easily visible, even those that lie in the vicinity of the Sun. Here and there, the Moon's edge is punctuated by the deep red prominences of the Sun's chromosphere, and the delicate pearly streamers of the corona, its outer atmosphere, spread outwards from the Sun. Another solar eclipse spectacle is Baily's Beads, a phenomenon caused by shafts of sunlight shining though lunar valleys at its limb, which gives the impression of a diamond-studded necklace. Since the Moon's umbral shadow barely touches the Earth – its diameter at the Earth's surface is typically 100 miles wide – totality is only visible from a small part of the Earth's surface, and totality never lasts more than 7 minutes and 40 seconds.

Lunar Eclipses

Eclipses of the Moon can only take place when the full Moon is situated less than 12.5° from one of its orbital nodes. There are three types of lunar eclipse: penumbral, partial and total. If the Moon traverses the outer penumbral shadow of the Earth, avoiding the umbra, a penumbral eclipse occurs. The penumbral shadow has no perceptible color, appearing gray with an outer edge that is impossible to detect visually. Unspectacular events, penumbral eclipses may not darken the full Moon to any noticeable extent. Partial lunar eclipses are far more satisfying to observe than penumbral eclipses. The Moon first enters the penumbra completely and then partially enters the umbra. The appearance of the umbral shadow is usually unmistakable. Partial lunar eclipses are designated a magnitude value from 0 to 1 for their maximum phase; for example, if the magnitude is 0.5, then half of the Moon's diameter is covered by the umbra at maximum eclipse.

The sharpness of the edge of the umbra varies a little from eclipse to eclipse. Sometimes the umbra fades gradually into the penumbra, but at other times it can appear sharp and unfuzzy. The umbra may be noticeably colored, but if the eclipse is partial and of a small magnitude, the glare of the uneclipsed portion of the Moon will make estimations of umbral hues and tones difficult. Many photographs of partial lunar eclipses show little detail or color within the umbra, since the exposure is kept short enough to prevent glare from the uneclipsed part of the Moon, but not long enough to register detail in the eclipsed part.

During a total lunar eclipse, the entire Moon enters the umbral shadow. The maximum possible magnitude of a total lunar eclipse is 1.888, meaning that the edge of the Moon farthest from the edge of the umbra is 1.888 lunar diameters away from the edge of the umbra at mideclipse. Lunar eclipses can last nearly

2 hours – a generous period in which to appreciate one of astronomy's most beautiful spectacles. Sunlight refracted by the Earth's atmosphere infiltrates the umbral shadow, so the Moon rarely totally disappears from view. Viewed from a totally eclipsed part of the Moon, the Earth appears as a large black circle surrounded by a ring of bright refracted sunlight, chiefly red in color.

Lunar eclipses are not entirely predictable; the hues, color distribution and intensity of the umbra always vary. Much depends on the magnitude of the eclipse, in addition to cloud and high-altitude dust in our own atmosphere, which affects the intensity of the sunlight refracted onto the Moon's surface. There appears to be a vague link between solar minimum and the intensity and redness of the umbra, but the cause of this is unknown. A more definite and predictable relationship exists between the darkness of lunar eclipses and the amount of dust released into the Earth's atmosphere by volcanoes. Vast amounts of dust thrown high into the atmosphere by the eruption of Krakatoa in August 1883 appear to have caused the darkness of the lunar eclipses of October 1884 and September 1888. More recently, the darkness of December 1992's lunar eclipse was likely to have been produced by the vast quantities of volcanic dust released into the upper atmosphere by the eruption of Mount Pinatubo in the Philippines in June 1991. The dust from the 20th Century's second biggest volcanic eruption quickly spread around the world, and colorful sunsets seen from Britain to Australia were attributed to it.

Lunar Occultations

When a distant celestial object, be it a star, planet or deep sky object, is temporarily hidden from view by the Moon, a lunar occultation takes place. The Moon has no appreciable atmosphere, and the stars are so distant that they appear as point sources of light. As a star is approached by the edge of the Moon, no fading effects are observed – the disappearance and reappearance of stars at the edge of the Moon occur nearly instantaneously when observed through the telescope. Planets move in paths close to the ecliptic, and they are occasionally occulted by the Moon. Unlike stars, planets have appreciable diameters, and they take a short period of time to be occulted: an average occultation of Jupiter at the leading limb of the Moon takes around a minute and a half.

With an apparent angular diameter of around half a degree, and moving against the starry background at a rate of around 13° each day, the Moon occults a total sky area of around 191.8 square degrees every lunar month, less than one two-hundredth the total area of the celestial sphere. Because the Moon's orbit is inclined by about 5° to the ecliptic, the area of sky prone to being hidden by the lunar disc lies exclusively within a band of 5° on either side of the ecliptic. Over a period of 18.61 years, the Moon occults all the stars in the 10°-wide zodiacal band; this amounts to around 1 in 10 of all the stars in the sky. Only 100 or so of these stars are near enough and large enough to have their actual diameters measured by occultation studies. The brightest of these, Aldebaran in Taurus, is 65 light years distant; it is an irregular, variable star that ranges from magnitudes 0.78 to 0.93. Occultation studies that accurately measure the drop in light at occultation have revealed Aldebaran to have a diameter of 20 milliseconds of arc – the apparent diameter of a star 50 km away.

Stellar occultation studies are important because they are an accurate means of determining the Moon's motion with respect to the background stars. Timings of occultations provides one of the parameters for monitoring Terrestrial Dynamical Time (TDT), the time scale used for calculating ephemerides of celestial events. Accurate occultation timings can reveal small variations in the Earth's axial rotation rate. Occultation timings may bring to light errors in a star's accepted position on the celestial sphere, or lead to a revision of the star's correct motion in the sky. Fading or staggered effects in an occulted star's light may suggest the presence of a previously unsuspected companion (or companions) to that star. Such observations are best made during grazing occultations, when the limb of the Moon just touches the star at maximum occultation and the star appears to graze along at a tangent to the lunar disc. The edge of the Moon is quite irregular, and grazing stars can appear to blink on and off several times as they disappear behind hills and mountains and reappear in lunar valleys.

The Search for Other Natural Satellites

Certain areas lying along the orbit of the Moon are capable of being occupied by a small object (or groups of them) without fear of rapid gravitational disruption into alternative orbits. These special sites are known as Lagrangian points, and they exist at 60° preceding and following the Moon, as measured from the Earth. Lagrangian points are not exclusive to the Moon: all large planet-sized bodies have them. Jupiter's Lagrangian points, for example, are occupied by the Trojan asteroid groups that orbit the Sun at the same distance and in the same plane as Jupiter, but 60° preceding and following the giant planet.

Telescopic searches for objects that might lie within the Moon's Lagrangian points have so far failed to reveal any signs of individual moonlets. If they were indeed present, moonlets as small as 20 m in diameter of the same albedo as the Moon would appear as bright as 12th-magnitude streaks of light (when fully illuminated) on photographs and CCD images. It is possible that much smaller moonlets exist, but they would have to be basketball-sized or smaller to have evaded detection for so long. Alternatively, it has been suggested that there might instead be large, diffuse clouds of material composed of billions of tiny particles ranging from a centimeter to less than a micron in diameter huddled in the vicinity of the Moon's Lagrangian points. These clouds could perhaps occupy a volume of space up to 50 times that of the Moon itself. Over the years, there have been occasional claims of the discovery of such wraithlike companions of the Moon, both photographically and visually, but none of these claims have stood up to rigorous examination. Advanced searches, made with large wide-field telescopes under ideal conditions, have consistently failed to reveal any trace of such ghostly clouds. Moreover, it is known that such insubstantial features, should they ever exist, would not remain for very long – the material would disperse into space just as quickly as it had accumulated.

Photographic surveys of near-Moon space have also ruled out the possibility that the Moon itself is attended by any appreciable natural satellites of its own. Searches have been based on the microscopic scrutiny of photographic plates. Dedicated searches have been made with large professional telescopes as well. A lunar satellite with a circular orbit of 46,700 km from the Moon will complete one lunar orbit, at an average velocity of 1,150 km/hour, in just over 11 days. The practical visibility of any satellite of the Moon depends upon the object's size, its reflective properties (albedo) and its proximity to the Moon's glare. The latter factor may be greatly reduced if the search were to take place during a total lunar eclipse, when the Moon was immersed in the Earth's umbral shadow. Searches conducted under these circumstances have found nothing of relevance, and we can be fairly certain that no object larger than 30 m (this rather large size assumes an albedo as low as the Moon's) exists in orbit around the Moon. Near-Moon space is undoubtedly packed with billions of tiny moonlets, ranging from tiny dust grains to sizable meteoroids, each cruising along its own doomed little lunar orbit. The supply of these objects is constantly being replenished as the Moon's gravity sweeps up material, much of it micrometeoroidal debris left in the wake of comets.

Chapter 3

Worlds in Comparison

We have seen that our satellite hasn't always been the placid orb that is implied by the names of some of its features, such as the Sea of Tranquility, Lake of Joy and Marsh of Sleep. Over the eons, innumerable meteoroids, asteroids and cometary nuclei have collided with the Moon. Working in conjunction with volcanic activity, these random impacts have violently sculpted the majestic lunar surface at which we now marvel through our telescope viewer.

Other solid worlds in the Solar System have been subjected to such titanic shaping forces. Some of these objects are blatant advertisements for the power of asteroidal impact, while others display bountiful evidence of having been extensively modeled from within by igneous or heat-related activity. Many planets and planetary satellites bear witness to both these processes in various proportions. An elite few of the Solar System's solid worlds have thick atmospheres capable of adding new dimensions to their surface topography by wind denudation, water erosion and sedimentation. Only one world we know of – our own Earth – has a dynamic biosphere. Life has had an immense effect upon our planet's external appearance, and human activity now has the potential to trigger a more vigorous, and potentially catastrophic, pace of change.

Many features on the four terrestrial planets – Mercury, Venus, Earth and Mars – bear striking resemblances to lunar features. However, there are no obvious lunar counterparts to features such as the Earth's midoceanic ridges, Venus's sprawling mountain plateaus and the circumpolar terraces of Mars. Precluded from our review are the Solar System's four largest planets – Jupiter, Saturn, Uranus and Neptune; these are all gas giants and have no discernible solid surfaces.

Mercury

With a diameter of 4,880 km, the innermost planet is somewhat larger than the Moon. At its largest, Mercury's apparent diameter reaches 12.9 arcseconds, equal in apparent size to the lunar crater Menelaus. Mercury's surface is full of craters, which have dramatic similarities to the Moon's highland regions. Closer scrutiny will reveal that more than a dozen different types of terrain exist on Mercury. The oldest is composed of highly eroded ancient craters and basins, interspersed with younger cratered areas, smooth plains, hummocky plains, ridges, escarpments and fractures. All these types of features can be found on the Moon. The range of Mercurian crater sizez and their different types closely match those of the Moon. There are small bowl-shaped craterlets, larger craters with central peaks, terracing

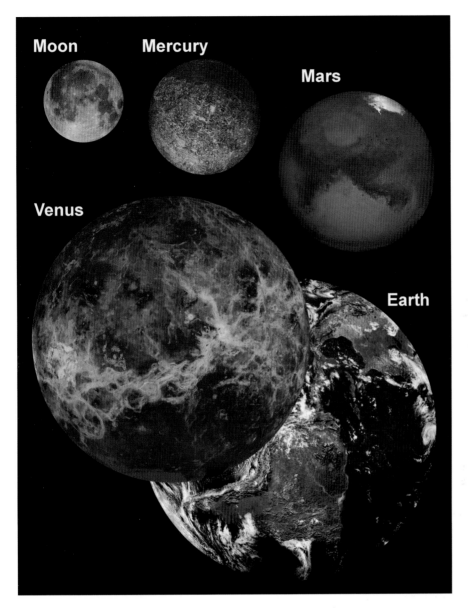

Fig. 3.1. Moon and terrestrial planets to scale.

and ray systems, and numerous dark-floored plains. Most of these craters were excavated by meteoritic impact, perhaps between 4.5 and 3 billion years ago. The ray systems and ejecta blankets tend to cover a smaller area in relation to crater size than their lunar counterparts because Mercury's gravitational pull is more than twice the Moon's. A conspicuous absence of large marial plains is the most significant difference between Mercury and the Moon. Mercury's single largest feature is Caloris Planitia, a basin 1,300 km in diameter, slightly larger than the Moon's Mare Imbrium. Caloris, so named because it is situated in the hottest area on the planet, is an asteroidal impact basin. The shock waves of the impact set up concentric crustal faults that became steep scarps and ridges as the crust adjusted to the stresses. Radial to Caloris are numerous ridges and valleys stretching for hundreds of kilometers. After the impact, there was a partial infilling of the basin by lava. A visual comparison between Caloris and the lunar far-side basin of Mare Orientale is quite striking.

There is more to the story of Mercury. Its metallic core is far bigger than the Moon's, perhaps in excess of 70% the planet's diameter. As the core cooled down after Mercury's formation, it shrank in size. Subsequent crustal adjustments and a crustal shrinkage of a few kilometers led to the formation of odd-looking wrinkles. Although topographically similar, the lunar wrinkle ridges (dorsa) are far smaller and less pronounced than those on Mercury, and they occur only in the lunar maria. The Moon's dorsa are largely a product of marial contraction.

Being the closest planet to the Sun, Mercury's sunlit face becomes incredibly hot, and it had always been assumed that there was no way that Mercury could ever host icy deposits. But in August 1991 a team of astronomers at the California Institute of Technology made an excellent series of radar images of Mercury's surface, using a high-power beam sent by the Jet Propulsion Laboratory's Goldstone antenna in California. Having received the radar reflections at New Mexico's Very Large Array radio telescope, the data was compiled into a map. Not only was the rough, cratered Mercurian surface revealed, but strong radar echoes were detected from the planet's north polar region, which happened to be tilted strongly towards the Earth at the time the observations were made. This clear radar return, imaged as a single brilliant spot around 400 km in diameter, has been interpreted as evidence for the existence of a large water ice deposit, just as radar (from lunar orbit) provided evidence for the lunar polar ice deposits. Like the lunar ice deposits, Mercury's polar ice (if it exists) probably arrived in icy cometary nuclei impacts and lies trapped within deep, permanently shadow-filled craters.

Venus

Venus's surface is permanently hidden from the eyes of visual observers because of the planet's thick cloud layers. Fortunately, radar (from the Earth and from probes orbiting Venus) has provided an excellent means of discerning surface detail. Early radar maps showed brighter, more radar-reflective areas that were thought to be hilly or mountainous regions, while the darker areas were thought to be smooth plains. One thing was evident – Venus did not possess anything like the Moon's circular marial basins.

Venus has shown two distinct types of terrain. Much of the planet is covered by an undulating landscape of low relief, a monotonous terrain accounting for around 90% of the surface. There are three major upland plateaus. The largest, Aphrodite Terra, has an area about equal to the entire Moon's. This huge mountain region stretches halfway around the planet south of the equator, and occasionally rises to over 7 km. The plateau is split by the enormous trench of Diana Chasma, which is 280 km wide and 4 km deep in places. In comparison, the Moon's Vallis Alpes (Alpine Valley) is a mere minor rift. Ishtar Terra, in Venus's northern latitudes, is a little smaller than Aphrodite, and boasts the planet's highest mountain, the 11-km-high Maxwell Montes, more than 3,000 m higher than the loftiest lunar mountains.

Venus's sprawling mountain continents appear to be the result of hundreds of millions of years of intensive vulcanism in the absence of plate tectonics and plate margin activity. Hot spots in the mantle have punctured the planet's relatively static crust and allowed huge accumulations of material to extrude onto the surface and sprawl across the local terrain.

Some of the smaller features on Venus resemble parts of the Moon. There are plenty of craters dotted at random about the planet, with central peaks, terraced walls and systems of debris surrounding them. Some craters sit atop high elevations, including pronounced, rounded swellings in Venus's crust, much like the summit craterlets on the lunar domes. There are several examples in which these domes appear to have collapsed, forming substantial sunken pits without raised rims, without comparison on the Moon. Parts of the planet are striated by parallel linear rilles comparable to the rilles found in many parts of the Moon.

Earth

In comparing the Earth with the Moon from afar, it would appear that no two bodies could be so entirely dissimilar in history, composition, form and appearance. Appearances can be misleading: studied up close, many of Earth's features possess topographic and structural likenesses and have near-identical modes of formation to those of the Moon.

The Earth has been on the receiving end of countless meteoritic and asteroidal impacts throughout its history. Because it was a larger target, the bombardment in Earth's early history must have been more intense than that on the Moon. It is estimated that during the past 600 million years, some 2,000 major impact events have occurred on the Earth. Yet even this phase must be regarded as quiescent compared with the cosmic catastrophes of several thousands of millions of years ago. The topographic evidence for ancient bombardment has largely been obliterated by the powerful forces of plate tectonics and vulcanism, in conjunction with the dynamic atmosphere, hydrosphere and biosphere. Not surprisingly, only a few impact scars can be identified with any certainty today.

Ancient impacts must have had a significant influence on the Earth's development. Global maps would be dramatically different had the Earth escaped repeated major cosmic assaults. An impact of the same magnitude as the one that excavated the Moon's Imbrium basin around 3 billion years ago would have easily shattered the young Earth's thin crust and stimulated large-scale vulcanism. Such an event would have been sufficient to have triggered a phase of tectonic activity and affected crustal dynamics enough to have changed the eventual configuration of entire plates.

Our knowledge of the Earth as it was 4 billion years ago is practically nonexistent: we shall never discover how the face of the Earth was sculpted by external forces further back than this time. It is sobering to realize that many of the lunar features visible through the telescope have been preserved intact from eons before the sketchy knowledge of our own planet's history begins.

Looking at a map of the Earth, it isn't difficult to identify outlines that suggest the highly eroded rims of impact craters. One of the largest and most prominent of these is on the eastern shore of Hudson Bay in Canada. This marks the edge of an ancient crater some 450 km in diameter; the central mountains of this feature are clearly marked by the Belcher Islands. There are many other bays and coves that, although seeming to suggest impact sites, have actually been formed by perfectly legitimate terrestrial processes. For example, it has been suggested that the Gulf of Mexico marks the northern periphery of a major ancient crater, as large as the lunar South Pole–Aitken basin. The hard geological evidence, however, does not support this idea.

The first terrestrial impact crater to have been discovered, and by far the most famous of its kind, was the Barringer "Meteor" crater in Arizona. At 1.3 km across, 175 m deep, with a rim raised over 40 m above the surrounding landscape, the Barringer crater is not large by lunar standards: if transplanted onto the Moon, it would be a minor craterlet that would require at least a 300mm telescope to be discerned from the Earth. It is estimated to have been formed around 27,000 years ago by an iron–nickel body some 70 metres across and weighing as much as 2 million tons. Humans across the world undoubtedly experienced the effects of the impact.

On the other side of the Earth, at Wolf Creek in Western Australia, lies another beautifully proportioned impact crater, discovered by oil prospectors in 1947. Somewhat smaller than the Barringer crater, the Wolf Creek crater is only 859 m across with a floor 30 m deep. It predates the Barringer formation by at least 20,000 years but, like Barringer, fragments of the original impactor have been found around the site.

Australia is home to a number of other impact craters. The famous Henbury crater cluster, just south of Alice Springs, encompasses an oval area of some 1.25 sq km. The largest of the group is shaped like the lunar crater Messier, but is far smaller, measuring 220 × 110 m with a floor up to 15 m deep. Because these craters lie in close proximity, the impacting body must have disintegrated at a low altitude and ploughed into the landscape from a southwesterly direction. The impact may have happened within the last 5,000 years. If so, it would have made a deep impression upon the region's early inhabitants: aboriginal legend tells us that the site of the Henbury craters is called "sun walk fire devil rock."

Around 65 million years ago, dinosaurs of the late Cretaceous period may have witnessed a significant asteroidal impact that devastated the (now) Yucatan peninsula area. Traces of a vast 200 km impact crater, buried hundreds of meters underground, were first detected in a series of aerial magnetic surveys. Current theories about the extinction of many dinosaur species credit this crater with having just the right age, size and location to have been responsible for this catastrophe. The crater – named Chicxulub, after a town in Mexico which overlies the site – is comparable in size to the lunar crater Clavius.

The tally of terrestrial impact craters is rising because of the increasing sophistication of geological investigation methods. More than 100 known astroblemes are known to be dotted about the Earth's surface, and no doubt many more will be uncovered in the years to come.

Mars

The red planet has a diameter of 6,790 km – more than twice the diameter of the Moon. Its apparent telescopic size varies between as much as 3.5 to 25.7 arcseconds, comparable to the apparent size of the tiny lunar crater Banting and the moderately sized crater Reinhold.

Like the Moon, Mars has two distinctly different-looking hemispheres. Mars's northern and southern hemispheres differ markedly in appearance. The southern hemisphere is covered with many lunar-type craters and on average is 3 km higher than the northern hemisphere. The northern hemisphere has undergone extensive vulcanism, and smooth lava plains have spread over the older crust, which many sizable volcanic shields have grown to tower over their surroundings. The dark areas of Mars mainly occupy its southern hemisphere in a near-continuous tract encircling the planet. Many of the darker regions project pointedly northwards. Syrtis Major is the most prominent of such features. Solis Lacus, the so-called "eye of Mars," is an interesting dark patch situated in a brighter circular region, outlined by a dusky crescent. Mare Acidalium is the most prominent of the far-northern hemisphere's dusky tracts. There are many brighter features competing for space on the Martian globe, including Hellas, Argyre and Elysium. The dark, mottled areas of Mars do not always conform to surface topography – even Syrtis Major's shape does not correspond to the lay of the land. The dark material from which many of the dark areas are composed is often found to emanate from small craters, and is blown by the Martian winds across local surfaces. Changes in tone and shape correspond to the variation of seasonal winds. The outlines of certain areas do reflect underlying topography. The bright areas of Hellas and Argyre mark the circular sites of ancient impact basins and are of the same size as the Moon's Mare Imbrium and Mare Orientale. The interiors of these prominent Martian regions have been filled to the brim with wind-deposited sand.

Asteroidal impact has produced the majority of large Martian craters; millions of years of subsequent weathering has modified their appearance. The steep interior slopes often possess marked gullies similar to those seen in the Earth's Barringer crater, and were probably cut by running water. The crater Arandas in Acidalia Planitia looks remarkably like the lunar crater Alpetragius, with a similarly huge central massif. What makes Arandas and many other Martian craters interesting is the presence of peculiar "mud-splash" patterns in the surrounding terrain, suggesting an impact on wet or frozen ground. Another unique feature is the streamlining of landscape around craters and other areas of pronounced relief, which indicates extensive water flow in Mars's history. Needless to say, unique formations like these are nowhere to be found on the Moon.

The Tharsis area is the most spectacular of Mars's volcanic regions, boasting four major volcanic shields that sprawl across the landscape. The largest of them is Olympus Mons, a giant mountain with a base 500 km across and a height of 24 km. To give an idea of its immensity, if transplanted onto the Moon, the volcano would easily cover Mare Humorum. Its lofty summit would catch the sunlight some 14 hours before local sunrise.

Vallis Marineris, to the southeast of Tharsis, is a vast canyon produced by crustal rifting. It is almost 5,000 km long, and, if placed on the Moon would stretch from lunar pole to pole. Comparable lunar valleys, like Vallis Alpes and Vallis Schrödinger, are dwarfed in comparison. Other Mars–Moon similarities include linear clefts and sinuous channels, although most of the latter were probably cut by running water, a phenomenon that has never occurred on the Moon.

The Satellites of Mars

Mars' two satellites, Phobos and Deimos, are tiny objects. Both are rocky, irregularly shaped and covered with craters. Phobos measures $27 \times 23 \times 19$ km, and Deimos measures $16 \times 11 \times 10$ km. Both have surfaces as dark as tarmac, reflecting a mere 4% of sunlight. Phobos is dominated by Stickney, a huge, deep crater that is no less than a third of the satellite's average diameter. Compared with the lunar crater Draper in the Mare Nubium, it is about the same size as Stickney. Phobos's surface is striated with parallel grooves, which, on closer inspection, are seen to be composed of interconnected crater chains with diameters up to a few hundred meters. The origin of these strange markings is undoubtedly linked to the formation of Stickney. It is possible that they were excavated by small, gaseous eruptions immediately after the Stickney impact, when water ice in Phobos's interior was superheated and exploded onto the surface. Crater chains of similar appearance do exist on the Moon, but in most cases they are chains of craters formed by the impact of material ejected by major impacts.

Deimos, like Phobos, has a pockmarked surface. Its surface is far smoother than its sister satellite's, and appears to be covered with fine-grained soil and larger boulders up to several tens of meters deep. This type of blanketing is not seem to a large degree on the Moon, although there are areas visible on the lunar surface covered by debris of a similar texture. It has been suggested that Deimos escaped the huge battering taken by Phobos and consequently had plenty of time to accumulate a meteoroid "sandblasted" regolith.

Asteroids and Comets

Every asteroid studied at close range has been found to be pitted with impact craters, some of them measuring a substantial proportion of the diameter of the asteroids themselves. Asteroid 951 Gaspra is a rocky body measuring $16 \times 14 \times 12$ km (about the same size as the Martian satellite Phobos). Its volume is about equal to the interior of the lunar crater Copernicus. Gaspra is covered with small impact craters, some showing minute ejecta systems. Grooves running across the asteroid's surface may have resulted from a large impact in the past. Interestingly, this diminutive body is thought to possess a marked magnetic field stronger than any on the Moon's surface. Asteroid 243 Ida, a cratered rock 52 km long, shows a long history of meteoroidal bombardment – many generations of craters are visible. Like Gaspra, Ida displays grooves indicating that it was once part of a larger body that experienced a catastrophic fragmentation. Near-Earth asteroid 4179 Toutatis is a bilobed body around 7 km across. The surface is cratered, with the largest crater measuring around 1 km. Asteroid 253 Mathilde is a minor planet that is more than twice as dark as the dark lunar plains. Mathilde, 59×47 km in diameter, has a massively cratered surface. The pitch-black asteroid is dented with five craters larger than 20 km across, the largest of which is more than three-quarters the asteroid's diameter and around 10 km deep! Proportionally, this crater is the largest impact feature on any body in the Solar System, proportionately bigger even than crater Stickney on the Martian moon Phobos or the South Pole–Aitken basin on the Moon. Just how Mathilde survived intact after experiencing such a battering remains a mystery; by any reasoning, it should have been smashed to bits long ago. The minor planet Vesta, 480 km across, is covered with dark and light features, but it has yet to be imaged up close. It is thought that Vesta's surface has been shaped by both impact and volcanic forces, much like our own Moon's.

The nucleus of Halley's Comet is an irregular pear-shaped lump of dirty ice measuring 16×8 km. Its surface is covered with shallow depressions, craters, hills, mountains, ridges and terraces. One crater, situated between two bright active areas, was observed to have a sharp rim and apparent central elevation. A good comparison of the actual size and shape of Halley's icy nucleus may be made with the elongated crater Messier in Mare Fecunditatis. By coincidence, the craters Messier and Messier A are known as the "comet" twins because of their elongated rays, stretching westward over the marial surface for over 100 km.

Jupiter's Satellites

There are 16 known satellites of Jupiter, and four of them – the Galilean satellites – are very large. Ganymede and Callisto are bigger than the planet Mercury, with diameters of 5,216 and 4,890 km, respectively. Io has a diameter of 3,636 km, just larger than our Moon. Europa is the smallest Galilean, with a diameter of 3,130 km. The other Jovian satellites are comparatively small, ranging from the irregular Amalthea (270 × 171 × 150 km) to the minuscule Leda, 15 km across.

Ganymede has no appreciable atmosphere. It is a world composed mainly of ice and carbonaceous compounds and has a reflectivity several times better than that of the lunar surface. Ganymede has areas of ancient, heavily cratered, crust, interspersed with regions that have been heated and remodeled by internal activity. Like the Moon, Ganymede has been the target of many meteoroids and asteroids. The icy nature of Ganymede's surface has meant that ancient impact scars have not been preserved, nor have the lunar features. Some of the larger impacts on Ganymede are clearly outlined by bright, concentric crustal furrows and moun-

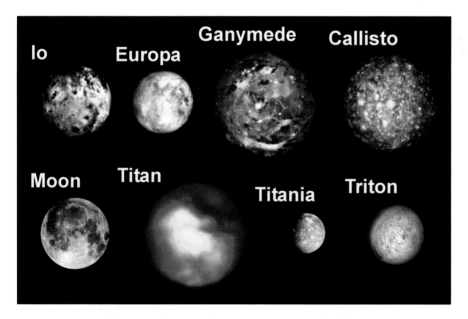

Fig. 3.2. The Moon compared with other big satellites.

tains, directly analogous to the fault systems surrounding the lunar marial basins. Smaller features include impact craters possessing central elevations and ray systems like those seen on the Moon.

Callisto is similar in appearance to Ganymede, with the same general composition, though it has a greater proportion of lighter ices and less rocky material. Its surface is dotted with brilliant craters, ejecta systems and concentric fault rings. Because Callisto is an icy world, a degree of mass material flow has taken place, and this has leveled out the big impact basins. Big impact craters have filled out, resulting in a fairly flat general relief. Internal heating and surface-deforming activity is believed to have been negligible in Callisto's history, and there was certainly no volcanic activity.

Io is a rocky world, composed of silicates and sulfur. It is the only planetary satellite currently in the clutches of active vulcanism, and its surface features are the newest in the entire Solar System. Active volcanoes on Io shoot material in vast plumes up to 300 km above the satellite's surface. Tidal pumping is responsible for Io's volcanic activity. Huge internal tides raised by the combined gravitational pulls of Jupiter, Europa and Ganymede lead to periodic frictional heating of Io's mantle. The crust of Io is estimated to be subjected to a tidal bulge of 100 m – a 100 times that created on the Earth by the Moon.

At first inspection, the surface of Io appears totally unlike that of the Moon. Large asteroids have undoubtedly hit Io many times in the past, but the scars of impact have long since been obliterated. Io's volcanic calderas are roughly circular in outline, and at their extremities are fringed with multicolored collars of ejected material. There are no radial ray systems, mountain chains or chain craters on Io, as the surface is in constant turmoil.

Europa is an icy world. The proportion of surface ice to silicates is greater than that of either Ganymede's or Callisto's. Europa's mean density suggests that its icy surface extends only a 100 km beneath the surface, the rest being composed of silicates. Europa is the smoothest, most topographically bland satellite in the Solar System: its features are comparable in magnitude to pencil markings on a pool ball. Europa is covered with thin, kilometer-wide clefts several hundred meters deep, somewhat like the lunar rilles. Some clefts, such as Asterius Linea, traverse the surface for many hundreds of kilometers. There are no large crater-like punctuations on Europa's surface, and only three craters in excess of 20 km are known. This strangely monotonous scene suggests that Europa's icy surface is fresh, the ice having moved rapidly across any old impact features, engulfing them and eroding them as it spread. There is probably a layer of liquid water between the outer ice layer and the inner rocky core; there is a strong similarity between fissures in the Earth's sea ice and cracks in Europa's crust. There is some suggestion that Europa has icy crustal plates with peculiar "ice volcanoes," fueled by internal heating, that have spewed out warm slush flows instead of hot lava. Because Europa is warm inside, with subcrustal oceans of liquid water, there is an outside possibility that some form of primitive marine life may have developed in this lonely outpost.

All of Jupiter's other satellites are irregular and pockmarked by impact craters. It is unlikely that these bodies were formed in the Jovian vicinity like the Galileans; instead, they were plucked from their independent asteroidal orbits by Jupiter's gravity. High-resolution images of some of these objects are currently being obtained by the Galileo Orbiter, which began its survey of the giant planet in 1995.

Saturn's Satellites

Saturn has 17 appreciably large satellites. Titan is by far the largest, 5,150 km in diameter. Nothing of the satellite's surface can be seen from afar because it is wrapped in a thick, yellow blanket of nitrogen and methane. Atmospheric banding produces equatorial belts and polar collars. Titan's thick atmosphere may have largely been supplied by cometary and planetesimal impacts. Of the other Saturnian satellites, Rhea, Iapetus, Dione, Tethys, Enceladus and Mimas are spherical bodies, all of them composed mainly of water ice. Rhea has a diameter of 1,500 km, somewhat less than half the diameter of the Moon. Rhea has a very high reflectivity, and its surface has undergone heavy bombardment over a long period of time. Some craters are ancient and eroded, while others appear relatively young, with bright, sharp rims. There are several large craters in excess of 50 km, and a large proportion contain prominent central peak systems. Much of Rhea resembles the jumbled lunar southern uplands. Some areas of the surface are mottled with light, wispy streaks, superficially resembling lunar rays, but their origin is probably internal, having resulted from the intrusion of ice along fissures in the crust. Iapetus is Saturn's third-largest satellite, measuring 1,400 km in diameter. It varies in apparent brightness by 1.7 magnitudes at either side of elongation since half of its surface has better reflectivity than the other. Iapetus's leading side reflects only 5% of incident sunlight, while the other hemisphere reflects up to 50%. The dark-leading hemisphere is probably the result of dark material's being extruded onto its surface. A number of Iapetus's craters have dark floors, and the impacts that produced them must have exposed and melted darker material inside the icy mantle. Some ejecta systems, too, are darker than their surroundings.

Dione, 1,100 km across, is smaller than Rhea, but the two satellites are similar in appearance, with many impact craters and light streaks feathering their surfaces. Both Dione and Rhea have similar histories and compositions. Tethys has an interesting surface, full of impact craters and huge fractures, one of which stretches more than the satellite's 1,040 km diameter. This feature, named Ithaca Chasma, was probably formed by crustal tension, as were many of the lunar rilles. The crust of newborn Tethys solidified in advance of its interior; when the interior began to freeze, it expanded, and the crust experienced a mighty crack. Enceladus has the most varied surface relief of any Saturnian satellite, given the possible exception of Titan's as-yet-unknown surface. Many distinct kinds of terrain exist on Enceladus, most notable of which are extensive areas without any traces of cratering. Tethys, therefore, may have experienced considerable heating that simply melted all the overlying older scars. Perhaps these smooth regions are as young as 200 million years. Mimas, 390 km in diameter, is very densely cratered. One impact crater, Arthur, measures 150 km across, over one-third the satellite's diameter. The impact that produced Arthur must have come very close to totally destroying the satellite. In places Arthur is 10 km deep, with a steep internal wall and prominent central mountain 4 km high.

Uranus's Satellites

Uranus has 15 large satellites. Titania, at 1,590 km the largest of Uranus's satellites, has undergone considerable crustal tension and faulting. Large scarps wind across the satellite's icy surface, many are similar in size to the lunar Rupes Altai scarp. The faulting probably occurred after Titania was formed, when the consolidated crust collapsed because the interior could not support it. Craters up to 50 km in diameter abound. The satellite's very rough surface texture suggests that recent crustal heating and melting is unlikely to have taken place. Oberon, 1,530 km across, is composed of ice and rock. Its mottled surface has an overall reflectivity of about 20%, with patches of much darker material. It is covered with impact craters surrounded by bright halos and rays. Some of these craters have dark floors, like some of those on Iapetus. Umbriel is the darkest of Uranus' major satellites, with a reflectivity of only 15%. Its surface is highly marked with large impact craters. A bright 150-km-diameter ring has a reflectivity twice that of Umbriel's average, and has around the same brilliance as the inner wall of Aristarchus. With a diameter of 1,160 km, Ariel is only a third the size of the Moon. The brightest areas reflect up to 50% of sunlight, while the darkest areas reflect only 20%. Its surface has experienced much more dynamic change than those of Titania, Oberon and Umbriel. The early history of the satellite is told in the many eroded and deformed impact scars that dot its surface. There has been a complex history, following its formation, of impact, crustal melting and ice movement, along with extensive faulting. Many of Ariel's features, like crustal melts and glacial flows, have no analogy with anything seen on the lunar surface.

Of all the satellites of Uranus – indeed, of all the satellites yet seen upclose in the entire Solar System – the weirdest satellite must be Miranda. This strange little world, only 480 km in diameter, has had a complex history involving active tectonics, crustal melting and movement, glaciation and asteroidal impact. In some regions there exist very ancient, rounded impact scars. Other regions are heavily striated, with linear faults and rilles. Some portions of the icy crust are bisected by huge angular cliffs, the steepest of which rises up to a spectacular 18 km above its surroundings. An interesting form of terrain on Miranda, the "circus maximi," is made up of roughly circular formations consisting of concentric grooves and ridges. Nothing remotely comparable exists on the Moon.

Neptune's Satellites

Triton, 2,700 km in diameter, is 75% the size of the Moon. Triton has great variety in its terrain. There are areas of very smooth, featureless plains, a strange landscape called "cantaloupe" terrain, and ancient valleys that have undergone glaciation. Interestingly, no very sizable craters exist on Triton. Strange, dark patches appear to have been the sites of local crustal melting. Active geysers shoot material up to 20 km above Triton's surface and are blown askew by high-altitude winds.

Triton might once have been an independent planet that was gravitationally captured by Neptune. Like many other bodies, Triton has been subjected to considerable tidal pumping and internal heating, which have transformed its original topography. The satellite is thought to have a large, rocky core surrounded by ice layers.

Proteus, Neptune's innermost satellite, is a potato-shaped body with a length of 400 km. Its dull gray surface (albedo of just 6%) is highly cratered, with one very large impact basin resembling the lunar crater Deslandres both in size and in its advanced state of erosion. The satellite is crossed by numerous linear structures that may have resulted from crustal fractures caused by a major impact.

As we find from this comparative tour of the Solar System, many features that exist on a host of other worlds show great similarities to parts of the Moon's surface. In some cases, the similarity is just one of general appearance, but there are direct geological comparisons too, even on satellites whose surfaces are icy. However finite the number of processes that have sculpted each of the Solar System's worlds, the most surprising fact is that each of them is noticeably different, but few show such variety or are as majestic as our sister planet, the Moon.

Part 2

Observing the Moon

Chapter 4

Observing and
Recording the Moon

These days, large changes don't take place very frequently on the Moon. Active vulcanism last took place more than a billion years ago. All that remains of an internally active Moon is the odd rumble at the base of its thick crust, in its warm, solid mantle region more than 600 km beneath the lunar surface. The lunar crust has long succumbed to all its pent-up stresses, deforming and faulting according-ly. Although asteroidal impacts do take place from time to time on the Moon, the last major impact that produced a crater large enough to be visible through back-yard telescopes probably took place long before the dawn of human civilization. It is likely that no lunar feature visible through a backyard telescope has been formed since the invention of the telescope. Down to the last detail, the Moon that was viewed by Galileo through his tiny refractor in the early 17th Century is probably utterly indistinguishable from the Moon viewed through a 21st-Century telescope eyepiece.

Until the advent of photography in the 19th Century, drawing at the eyepiece was the only means of recording the Moon's features. Professional astronomers inter-ested in researching the Moon haven't needed to peer through telescope armed with a pencil and sketchpad for more than a century. Now the topography of the Moon's surface – both near side, far side and polar regions – has been mapped in great detail, and the composition of its surface is well known. These days, few professional observatories ever turn their big telescopes towards the Moon other than to impress new students or to test out new equipment. Observers of faint deep sky objects, both professional and amateur, regard the Moon as something of a nuisance, an inconvenient source of light pollution whose glare drowns out the faint light from distant nebulae and galaxies.

Lunar observing is almost entirely the province of the amateur astronomer. Anyone who looks at the Moon with a purpose can be called a lunar observer. That purpose may be to attempt to measure the changing apparent diameter of the Moon during the course of a lunation, using a simple naked eye cross-staff. Alternatively, the purpose may be to conduct painstaking research into, say, the changing appearance of low wrinkle ridges in Mare Humorum at the eyepiece of a very large telescope. Both of these activities, lying at opposite ends of the lunar observing spectrum, aim to answer scientific questions. But equally valid a pursuit is observing the Moon for the sheer visual pleasure it brings. Few people who have ever viewed the Moon through a telescope can have failed to be impressed by its grandeur; the constantly changing vistas of the Moon's surface are every bit as stimulating as the contemplation of an Impressionist painting. Many lunar observers will freely admit (I certainly do) that they often imagine their telescope eyepiece to be the porthole of their very own spacecraft hovering over the Moon's surface. Just seeing the Moon's surface through the telescope and being able to locate and identify the main lunar features is satisfying enough for many lunar observers. But many lunar observers want to take their enjoyment of the Moon a step further by making their own permanent record of their forays around the Moon's surface.

Drawing the Moon

Digicams, camcorders, webcams and dedicated astronomical charge-coupled device (CCD) cameras are capable of securing detailed images of the Moon, recording features that the visual observer would find impossible to draw with any degree of accuracy. So, what possible reason is there to draw the Moon's features, an activity that appears to belong to the distant past? Surely, the lunar surface appears so detailed, even through a small telescope, that only an expert artist could hope to accurately draw even a small section of it. What is the point, anyway, as CCD cameras can do it all effortlessly in a fraction of the time? Besides, why bother to draw or even image the Moon through a backyard telescope when the entire lunar surface has already been mapped in exquisite detail?

Such arguments are among those used nowadays to dismiss the efforts of the observer who sketches the Moon's features or, for that matter, makes an observational drawing of any other celestial object. These arguments completely miss the point of why many lunar observers choose to draw the Moon's surface. Learning how to draw the Moon's features is an activity that has the potential to improve every single aspect of the amateur astronomer's skills of observation.

With its hundreds of craters and mountains and vast gray plains, the lunar surface can appear utterly confusing to a novice, but it becomes increasingly familiar over time. Novices are often amazed at how the effects of illumination vary with the distance from the terminator: craters look incredibly deep near the terminator, but a little distance from it they may not look so cavernous. At first, the observer, armed with a map, may struggle to correlate the view through the telescope with the major features marked on a map. But, as the major landmarks are noted, they can be used as stepping stones to hop across to other, more subtle, features that may have originally been overlooked. Without exception, an observer's ability to discern the fine detail on the Moon will improve in proportion to the amount of time that is spent studying the lunar surface through the telescope eyepiece.

Just studying the Moon through the eyepiece with the aid of a map, photographs and/or a written description hones the skills of observing and is a continual learning experience. But one certain means of fine-tuning one's skills of observation that goes hand in hand with learning one's way around the Moon is to spend time drawing individual features or small groups of features. By paying full attention to a small area of the Moon, instead of allowing the eye to roam around the lunar surface, the brain begins to make sense out of the light and shade, teasing fine topographic detail from what may have initially looked like a plain hole in the ground. It is remarkable how much more detail can be discerned when the observer has the task of attempting to draw a single small feature as accurately as possible.

It is important to have confidence in your own drawing abilities. Drawing should be an enjoyable pursuit, but it is surprising how many people's experiences at school art class tend to have been negative one. If this is the case, you have my permission to disregard everything your art teacher ever told you! The lunar observer isn't some kind of weird nocturnal art student – nobody is allocating marks for artistic flair or the aesthetic appeal of an observational drawing. What is important is the effort that an observer puts into the attempt, and observational honesty and accuracy count above all. The aim of the exercise is to learn the topography of the Moon, to attend to the fine detail visible through the eyepiece and record it to the best of your abilities, not to produce an artistic masterpiece. The finished results will be the products of your efforts and a permanent record of your forays around the lunar surface. Don't throw them away – keep all your lunar observations in a folder, and you may be pleasantly surprised at how you improve over time.

Tonal Sketches

At the eyepiece, find your bearings with a reasonably detailed map of the Moon that is easy to handle at the telescope; a photocopied set of the lunar maps contained in this book may work very well. Select and identify your target, such as an individual crater, and one preferably close to the lunar terminator where most relief detail is visible (drawing features that are completely devoid of shadow is a specialist skill in itself; see below). If the feature that you have chosen to observe isn't marked on your map, make a note of any prominent features nearby that can aid you in identifying the feature later on. If possible, return indoors and make a basic light-pencil outline drawing of the features within your chosen area, using your map as a guide. Take care to get the proportions and arrangement of features right; this will save time and give you a distinct advantage at the eyepiece. Remember that features near the Moon's limb are affected by libration, and an outline copied from a map may not be the same as that feature on the evening that you have chosen to observe.

To make regular pencil sketches, a good range of soft-leaded pencils and a small pad of smooth cartridge paper are recommended. Basic outlines of lunar features are first drawn very lightly, using a soft pencil, giving you the chance to erase anything if the need arises. Scale is important – thumbnail sketches will not convey nearly enough detail, but an individual drawing that takes up the whole page will take far too long to complete to one's satisfaction. I have found that a drawing of between 75 and 100 mm diameter is quite sufficient for most observations. When shading dark areas, it is best to apply minimal pressure on the paper. The darkest areas of shadow are ideally applied in layers of soft pencil rather than by a frenzy of heavy pencil pressure.

It is reasonable to set yourself about an hour or two per observational drawing. Patience is essential, because a rushed sketch is bound to be inaccurate. Even if the clouds are threatening to obscure the Moon from view, or if your fingers are feeling numb with cold, it's best to have an accurate half-completed drawing rather than an inaccurate drawing of the entire area you have observed. Short, written notes can be made at the telescope eyepiece, pointing out any unusual or interesting features that you have observed, that may not necessarily be obvious on your

Fig. 4.1. Showing the stages in making an observational drawing in soft pencil. The feature is la Condamine. Credit: Peter Grego

drawing. Of course, it is essential to note the name of the feature you have observed, the date, start and finish times (in Universal Time, also called Greenwich Mean Time) of the observation, the instrument and magnification used, and the viewing conditions. Copies of your observational drawings ought to be made as soon as possible after the observing session, while the information remains fresh in your mind.

Sketching skills can be improved by drawing small sections of detailed lunar photographs that appear in books and magazines. After several attempts at "armchair" Moon drawing, you may surprise yourself at how quickly you improve. The most important thing is to be patient and not to rush, even if you are only practicing.

Line Drawing

An alternative to making shaded pencil drawings at the eyepiece is to represent features in outline form. Many observers choose to make annotated line drawings at the eyepiece, using the technique as a form of observational shorthand, along with intensity estimations (see below). The line drawings, and the written inform- ation that accompanies them, are then converted into full tonal renditions in pencil or ink after the observing session. However, line drawing should not be considered a quick and easy alternative to tonal drawing, since a line drawing ought to be drawn just as carefully, with the same amount of attention to detail.

Obvious features such as the rims of prominent craters and the sharp outlines of shadows cast by mountains are represented by bold lines. Try to avoid depicting lunar mountains as upturned V shapes – this might be fine for cartoon strips, but it only serves to confuse the appearance of a lunar observational drawing. If a part of the drawing depicts mountainous terrain, attempt to delineate the borders of the terrain, outlining the main peaks, features of relief and main shadows. Similarly, a detailed area of fairly homogenous rough terrain that contains perhaps too much fine detail to depict accurately ought not be drawn as a mass of dots and jagged squiggles: simply try to mark the borders of the terrain and label it "rough terrain." Subtler features, such as lunar domes and wrinkle ridges, may be recorded with lightly applied lines. Dashed lines can be used to delineate features like rays, and dotted lines can be used to mark the boundaries of areas of different tone.

Line drawings do require plenty of descriptive notes to accompany them, far more does than a well-executed tonal pencil drawing, which can stand on its own with the minimum of notes. The line-drawing method has the advantage of requiring a minimum of drawing ability, although observational accuracy is just as important as in tonal drawing. When used properly, the method can be as accurate and as full of information as any toned pencil drawing. Performed by someone used to the technique, line drawing can be quicker to accomplish than a tonal pencil sketch.

Intensity Estimates

A basic annotated line drawing can convey the appearance of a lunar feature and its topography in great detail. However, along the terminator the Moon's surface varies in brightness from pitch black, through all shades of gray, to dazzling white. It would be quite a laborious task, if not impossible in many cases, to go to the trouble of labeling each area of different tone on a line drawing in longhand. To make the task manageable, an intensity-estimates shorthand is used to accompany line drawings. This requires that the observer to estimate the brightness of each distinct area depicted on the drawing, using a scale of 0 to 10, 0 for the blackest lunar shadows, 10 for the most brilliant areas.

Intensity-Estimate Scale

The tonal examples given below are based on a general binocular or low-power telescopic view of areas that are illuminated by the midday Sun. Each area breaks down into further tonal grades when scrutinized at higher magnification, and each area will tend to appear darker the nearer it lies to the terminator.

0. Black –darkest lunar shadows
1. Very dark grayish black – dark features under extremely shallow illumination
2. Dark gray – the southern half of Grimaldi's floor
3. Medium gray – the northern half of Grimaldi's floor
4. Medium light gray – general tone of area west of Proclus
5. Pure light gray – general tone of Archimedes's floor
6. Light whitish gray – the ray system of Copernicus
7. Grayish white – the ray system of Kepler
8. Pure white – the southern floor of Copernicus
9. Glittering white – Tycho's rim
10. Brilliant white – the bright central peak of Aristarchus

The eye is capable of differentiating between hundreds of shades of gray, so a skilled observer can easily subdivide the basic scale yet further. Unlike variable-star estimations, these tend to be qualitative visual estimates rather than quantitative ones. For example, an individual mountain peak approaching the evening terminator may not appear particularly brilliant, having an intensity rating of perhaps 6 or 7. But once the dark shadows of lunar night spread around the landscape surrounding it, the mountain peak may appear like a dazzling beacon, with an intensity rating of 9 or 10 glinting beyond the terminator, even though the peak is in fact less bright than when the mountain lay in full sunlight.

Copying Up

Nobody has the ability to produce completely error-free observational drawings at the eyepiece, so it is best to prepare a neat copy of your observational drawing as soon after the observing session as is practically possible, while the scene remains vivid in your mind. A fresh drawing prepared indoors will, it is hoped, be far more accurate than the original telescopic sketch that it is based upon, as the observer will be able to recall little things about the original sketch that perhaps weren't quite right and needed to be rectified on the neat drawing. To improve the accuracy of a drawing, it can be useful to base the general outlines of a neat drawing upon an outline blank, a map or photograph of the area that has been observed. This is perfectly legitimate as long as the observer isn't adding any detail that wasn't actually observed. The neat copy of your original telescopic sketch can be used as the template for further drawings, or used to electronically scan or photocopy.

Copied drawings may be prepared in a variety of media. Superb results can be achieved using India ink washes and paintings in gouache – a watercolor medium that is possible to apply fairly thickly and in a controlled manner. Both are excellent for reproducing observations on a larger scale for exhibition purposes. Both these techniques require proficiency in brushwork, and although a description of the methods involved is beyond this book, practice, experimentation and perseverance will pay big dividends.

My own preference – soft pencil on smooth cartridge paper – is by far the quickest and least fussy medium. Once completed, pencil drawings need to be sprayed with a fixative so that they don't smudge if they are inadvertently rubbed. Regular copy shop photocopies of tonal pencil drawings are not quite good enough to submit to astronomical society observing sections or for publication in magazines, as the full range of tones in the drawing will not be captured, and they may appear somewhat dark and grainy. These days, however, most observing sections are happy to accept a high-quality laser print or a digitally scanned drawing. Some commercial magazines may insist on having the original artwork to work from or, at the very least, high-quality, high-resolution scans submitted on floppy disc or by email.

Ink stippling is one method of reproducing observational drawings that looks absolutely superb when done skillfully. Closely spaced dots of black ink are used to convey the illusion of shade – the darkness of the shade increases with more closely spaced and/or bigger dots. Areas of black shadow are simply blacked in with ink and a brush. Being composed of thousands of individually applied black dots, stippled drawings reproduce very nicely even when they are photocopied. Stippling must be performed using a set of technical pens with a range of nib sizes; an ordinary fountain pen, felt-tip or ballpoint pen simply isn't good enough. Stippling takes a great deal of time, and requires patience and an exceedingly steady hand. An individual capable of producing wonderful pencil drawings may not necessarily be able to master stippling. For example, my own efforts to get to grips with the technique have not been too successful. Stippling is an exacting process that does not easily forgive lapses of attention, and drawings do not welcome extensive alteration. A small group of dots inadvertently placed too close together may indicate a nonexistent feature. Stippling in too regimented a fashion – along straight lines, for example – may give rise to unwelcome artifacts. So, as well as having a disciplined eye and hand, there must be a certain relaxation of technique and an ability to keep the bigger picture in mind. In the world of lunar observation, there are only a handful of truly competent masters of the stippling technique.

Tempting though it may be to discard old drawings, they represent a permanent record of your observations and hard work at the eyepiece. At the very least, comparing old observations with more recent ones will demonstrate how much your skills of observation and recording have improved. Older observations can be compared with more recent observations of the same feature. Remember that constant changes in illumination and libration means that lunar features can appear markedly different from one hour to the next, as well as from one lunation to the next, and your drawings may have captured an aspect of a feature that is rarely observed. Original drawings can be used as the basis for subsequent copies for the observing sections of any astronomical societies to which you belong, or for publication. For these reasons, do hang on to your original drawings for future reference. Devote a folder or a ring-binder to your original lunar drawings and store them safely in a dry environment.

Observational Information

Finished observational drawings should ideally be accompanied by more than just the basic information of the feature name, date and instrument used. A complete observation may include some or all of the following details:

Feature Names

International Astronomical Union (IAU) official names of features should invariably be used. Many older atlases may contain errors of nomenclature or outmoded and unofficial designations for features. "Pickering," for example, was a name once unofficially given to Messier A in Mare Fecunditatis. However, the official IAU crater Pickering is a different crater, located near the center of the Moon's disk. If the feature that was observed cannot be identified with an official IAU designation, note its location in relation to several of the nearest named features – "small group of hills in Mare Frigoris, immediately west of Harpalus," for example.

Date and Time

Amateur astronomers around the world use Universal Time (UT), which is the same as Greenwich Mean Time (GMT). Observers need be aware of the time difference introduced by the time zone in which they reside, and any local daylight-savings adjustments to the time, and convert this to UT accordingly; the date should be adjusted too. Times are usually given in terms of a 24-hour clock – for example, 3.25 P.M. UT can be written as either 15:25, 1525 or 15h 25m UT.

Viewing and Scales

To estimate the quality of an astronomical view, astronomers refer to one of two visual scales. In the U.K., many observers use the **Antoniadi scale**, devised specifically for lunar and planetary observers.

AI – Perfect viewing, without a quiver. Maximum magnification can be used if desired

AII – Good viewing. Slight undulations, with moments of calm lasting several seconds

AIII – Moderate viewing, with large atmospheric tremors

AIV – Poor viewing, with constant troublesome undulations

AV–Very bad viewing. Image extremely unstable; hardly worth attempting to observe lunar surface features

In the United States, viewing is often measured from 1 to 10 on the **Pickering scale**. The scale was devised according to the appearance of a highly magnified star and its surrounding airy pattern through a small refractor. The airy pattern, an artifact introduced by optics, will distort according to the degree of atmospheric turbulence along its light path. Under perfect viewing conditions, stars look like tiny bright points surrounded by a complete set of perfect rings in constant view. Of course, most lunar observers don't check the airy pattern of stars each time they estimate the quality of viewing during an observation; an estimate is made based on the steadiness of the lunar image:

P1 – Terrible viewing. Star image is usually about twice the diameter of the third diffraction ring (if the ring could be seen)

P2 – Extremely poor viewing. Image occasionally twice the diameter of the third ring

P3 – Very poor viewing. Image about the same diameter as the third ring and brighter at the center

P4 – Poor viewing. The central disc often visible; arcs of diffraction rings sometimes seen

P5 – Moderate viewing. Disc always visible; arcs frequently seen

P6 – Moderate to good viewing. Disc always visible; short arcs constantly seen

P7 – Good viewing. Disc sometimes sharply defined; rings seen as long arcs or complete circles

P8 – Very good viewing. Disc always sharply defined; rings as long arcs or complete but in motion

P9 – Excellent viewing. Inner ring stationary. Outer rings momentarily stationary

P10 – Perfect viewing. Complete diffraction pattern is stationary

Considerable confusion can be caused if a simple figure is used to estimate viewing without indicating whether it's made on the Antoniadi or Pickering scale. So, in addition to designating the viewing with a letter and a figure (AI to AV or P1 to P10), a brief, written description of viewing, such as "AII – Good with occasional moments of excellent sight," can be made.

Conditions

These are indications of the prevailing weather conditions, such as the amount of cloud cover, the degree and direction of wind and the temperature.

Transparency

More relevant to deep-sky observers than lunar observers, the quality of atmospheric clarity is known as "transparency." Transparency varies according to the amount of smoke and dust particles in the atmosphere, as well as cloud and haze. Industrial and domestic pollution causes transparency to be worse in and around cities. A transparency scale of 1 to 6 is often used, according to the magnitude of the faintest star detectable with the unaided eye. Some lunar observers choose to include a measure of transparency for the sake of completeness.

Age and Phase

An average lunar month, or lunation, lasts 29d 12h 44m from one new Moon to the next. Many observers choose to note the time elapsed from that lunation's new Moon in days and hours to the time of the observation. The age of the Moon may be used to estimate the approximate phase of the Moon at the time of observation. The Moon's phase can also be indicated by a short note, such as "waxing gibbous" or "waning crescent." A more exact figure used to gauge the Moon's phase denotes the percentage of the near side that appears illuminated. For example, at first quarter phase, the Moon is 50% illuminated (waxing); at last quarter it is 50% illuminated (waning).

Libration

Combined, the effects of libration on latitude and longitude cause a displacement of the Moon's mean centre. Features on the Moon's mean far side are brought onto the Earth-turned lunar hemisphere on one limb, while features move onto the hemisphere facing away from the Earth on the opposite limb. At extremes, libration in latitude amounts to ±6.5°, while libration in longitude can be as great as ±7.5°. These figures are useful to quote, especially on observations of near-limb features in the libration zone.

Selenographic Coordinates

Coordinates on the Moon are termed "selenographic coordinates." The mean center of the lunar disc is located at a selenographic latitude of 0° and a selenographic longitude of 0°. Going northward, selenographic latitude is positive, and negative going southward. The Moon's north pole is located at a selenographic latitude of 90°, and the south pole is at –90°. Selenographic longitude increases eastwards from the Moon's mean central meridian. The mean eastern limb is located at 90°, increasing eastward around the far side to 180° at the point opposite the mean center of the disk, 270° at the mean western limb and 360° (0°) at the mean center of the disc. On maps, features either side of longitude 0° are often marked as positive (degrees east) or negative (degrees west). A feature located at +45° (45°E) corresponds to a selenographic longitude of 45°. A feature located at –45° (45°W) corresponds to a selenographic longitude of 315°. A feature's selenographic coordinates can be indicated on the observational drawing.

Selenographic Colongitude and the Lunar Terminator

If the Moon maintained a perfectly circular orbit around the Earth and did not undergo librations, it would present the same face towards the Earth: its morning and evening terminators would cross over the same features at exactly the same point in each and every lunation. But this is not the case, and libration has a considerable effect on the apparent position of features in relation to the terminator. For example, the eastern part of the large crater Ptolemaeus lies on the Moon's mean central meridian, at a selenographic longitude of 0°. Therefore, at a mean libration, Ptolemaeus appears on the sunrise terminator at precisely first quarter phase. However, at extremes of libration, Ptolemaeus can lie more than 3° east or west of the mean center of the Moon, and its appearance on the terminator can be advanced by more than 12 hours prior to first quarter or retarded by more than 12 hours after first quarter phase.

In order to calculate precisely where the morning or evening terminator of the Moon lies, a figure called the Sun's selenographic colongitude is used (see Table 4.1). The Sun's selenographic colongitude is numerically equal to the selenographic longitude of the morning terminator; the figure is published in astronomical ephemerides and can be displayed on many lunar computer programs. At new Moon, the Sun's selenographic colongitude equals 270°, at first quarter 90°, full Moon 180° and last quarter 180°. To work out exactly where the near-side morning or evening terminator lies in relation to the actual selenographic longitude at the Moon's equator, see Table 4.1.

The Sun's selenographic colongitude increases by approximately 0.5° per hour, or 12° per day.

Table 4.1. Calculating the position of the Moon's terminator

Phase	Terminator	Sun's selenographic colongitude (S)	Longitude of terminator
New Moon to first quarter	Morning	270° to 360°	360°-S East
First quarter to full Moon	Morning	0° to 90°	S West
Full Moon to last quarter	Evening	90° to 180°	180°-S East
Last quarter to new Moon	Evening	180° to 270°	S-180° West

Sun's Selenographic Latitude

Another figure that many lunar observers note on their observations is the Sun's selenographic latitude. This equals the Sun's selenographic latitude at the subsolar point, and varies between $+1.5°$ and $-1.5°$ over the course of around 6 months.

Lunation

The numbering of the Moon's lunations officially began with lunation 1 on 16 January 1923. Lunation 1,000 began on 25 October 2003. Many lunar observers include a lunation number on their observational reports.

Lunar Data

Much of the necessary information about the Moon is published in annual astronomical ephemerides such as the *Handbook of the British Astronomical Association*, the *Astronomical Almanac* (a joint publication of the U.S. Nautical Almanac Office and Her Majesty's Nautical Almanac Office in the U.K.) and the Multiyear Interactive Computer Almanac (MICA; published by the U.S. Naval Observatory). Programs for personal computers are becoming increasingly sophisticated, and are capable of giving the user much more than a few dry figures. Many of the more advanced lunar programs are capable of displaying high-resolution images or maps of the surface of the Moon, fully adjusted for phase and libration. It won't be long before personal computers are capable of running advanced programs that display detailed topographical three dimensional models of the Moon's surface, adjusted for lighting and libration, that show all the features that can be discerned through a small telescope. Such a program could only help encourage amateur astronomers to take out their telescopes and view the Moon's splendors with their own eyes.

Imaging the Moon

Conventional Photography

Even a small telescope can deliver a beautiful, pin-sharp image of the Moon's surface to the observer's eye, so it may seem reasonable to suppose that something of this can be captured on film without too much difficulty. Being such a large, bright object, the Moon will indeed register on just about any film camera pointed through a telescope eyepiece, and most amateur astronomers have experimented with basic lunar photography by doing just this, clicking away in the hope that a lunar image of sorts will be captured. However, successful lunar photography using an ordinary film camera – be it a simple compact camera or a 35mm SLR (single lens reflex) – is more involved than many imagine.

Viewed through an undriven telescope, when the Moon is drifting across the field of view, the observer can easily fix on an object. An experienced observer will

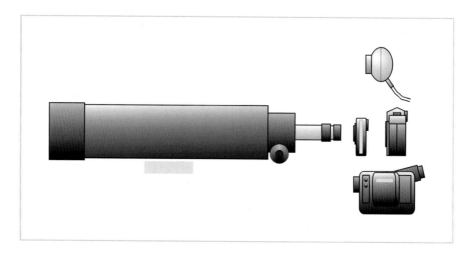

Fig. 4.2. Lunar imaging can be performed with CCDs/webcams (top), digicams/compact fixed-lens cameras and SLRs (middle) and camcorders (bottom). Credit: Peter Grego

be able to take advantage of moments of good visibility, appreciating what detail there is to be seen, so that it is possible to make lunar observations even in poor viewing conditions. A camera has none of these advantages. Whatever the exposure used to photograph the Moon, the effects of image drift in an undriven telescope will blur the resulting image to some extent. Hand-holding a camera to the telescope eyepiece will also introduce blurring. A camera needs to be fixed firmly to a driven telescope, keeping the Moon steady and centered in the field of view.

Afocal Photography

Afocal photography – photographing an object through the telescope eyepiece with the camera's lens also in place – can be performed with both conventional film cameras and digital cameras with fixed lenses. The most basic compact film and digital cameras have fixed lenses that are preset to focus onobjects from a few metres away to infinity; this focus cannot be varied. Additionally, the exposure may be preset for standard photography in daylight. The viewfinder of a basic camera may be a simple rectangular aperture with a small lens; this will of course be completely useless for afocal photography, as it will not show the magnified image being projected into the camera. Given that the view through the camera will not be visible during photography, the Moon must be kept precisely aligned through the cross hairs of an accurately aligned viewer.

Even with such limitations, afocal photography through such a basic camera can produce pleasing results. A low-magnification eyepiece will deliver a brighter image and reduce the effects of the Moon's drift through the field of view if the telescope is undriven. If there is a focus adjustment on the camera, it should be set to infinity. First, the Moon is focused by eye through the telescope eyepiece, and the camera is then positioned close to the eyepiece and held there firmly. Since most basic cameras are meant to be used as they are, without any fancy accessories, some kind of simple makeshift adapter will probably need to be constructed; some

Fig. 4.3. First quarter Moon, imaged afocally using 150mm refractor and Canon digital camera. Credit: Doug Daniels

cameras are so lightweight that some blu-tack and electrical tape will be sufficient to temporarily fix them to a telescope.

An eyepiece may deliver a large apparent field of view to the observer, and the dark edge of the field may be quite unobtrusive visually. But afocal images are particularly prone to vignetting, where the Moon is surrounded by a dark circular border, making it look as though the photograph had been taken through the porthole of a spacecraft. Indeed, some vignetting is noticeable in a number of lunar images taken professionally – even appearing on some photographs reproduced in the famous *Consolidated Lunar Atlas*. Vignetting mainly happens when the camera lens is much larger than the eyepiece lens. Eyepieces with narrow apparent fields of view are unsatisfactory; this includes most "budget" eyepieces such as Huygenian and Ramsden types. In addition, eyepieces with very short eye relief are difficult to use afocally with many fixed-lens cameras because both the camera and eyepiece lenses need to be positioned so close to each other that they are almost in direct contact.

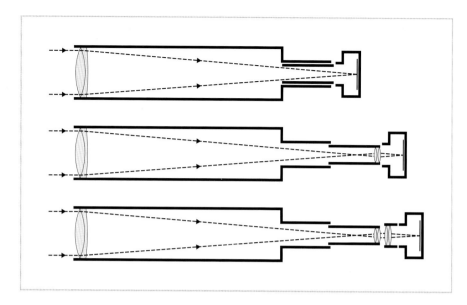

Fig. 4.4. The three methods of lunar photography at the telescope. Prime focus (top), eyepiece projection (middle) and afocal photography (bottom). Credit: Peter Grego

SLR Photography

SLR (single lens reflex) cameras direct the light from the object being photographed through the main camera lens to the eye via a mirror, prism and eyepiece. The area framed in the viewfinder is exactly the area that will be included in the final image. When the shutter button is pressed, the mirror instantly flips out of the way of the light path, allowing it to project directly onto the film. The ability to view the subject directly immediately prior to taking a photograph means that the SLR is perfectly suited for capturing an image of the Moon through the telescope. SLR cameras have grown increasingly sophisticated since their general introduction in the 1960s. Top-of-the-line SLRs are provided with electronic controls that can automatically adjust every aspect of the camera's functions. For lunar photography, however, a basic SLR with mainly manual controls can produce just as good a lunar photograph as a state-of-the-art computerized SLR. To minimize the effects of vibration that occurs when the shutter button is pressed, a cable release can be used. Better still, most SLRs can be adjusted to take an exposure after a timed delay, giving the vibrations in the telescope time to damp down after the camera is handled. In some basic SLRs, the movement of the shutter can vibrate the camera and telescope, causing a little blurring of the image; this would, of course, present more of a problem when taking high-magnification lunar images, as the effects of any movement of the telescope would be exacerbated.

Film Types

A film's ISO rating indicates its speed: the higher the ISO, the faster the film and the less exposure time is required. A medium-speed film is 200 ISO and inexpensive generic 200 ISO color print film is fine for beginning lunar photographers to experiment with. Quality tends to vary from brand to brand, and even varies among different batches of the same budget brand. Slower films have finer grains, allowing more detail to be captured, while grain size increases with a film's ISO rating. This may not be evident in comparing regular sized photographic prints, but enlargements will clearly show the difference. Photographs taken with slower film can withstand much more enlargement than those taken with faster film, but slow films also have the drawback of requiring more time to expose. High-magnification shots of the lunar surface made using slow film require that the camera be attached to a telephoto lens or telescope on a driven equatorial drive.

Telephoto Lenses

In its simplest form, a telephoto lens is a small, refracting telescope that can be attached to the front of an SLR camera. Telephoto lenses have a longer focal length than the lenses ordinarily supplied with SLRs: the longer their focal length, the higher the magnification they produce. Varifocal telephoto lenses contain an array of optical elements that allow its focal length to be increased and the image zoomed in without the need to adjust the camera's focus on some of the better models. There are a number of advantages to using a telephoto lens on an SLR to photograph the Moon. Since they are relatively lightweight and portable, telephoto lenses can be set up quickly and used to photograph the Moon from areas of the observing site that are inaccessible to a bulkier telescope and its mount. Provided that the camera and telephoto lens are held steady, photographs that show a considerable amount of detail along the Moon's terminator can be captured. Faster films (with higher ISO ratings) will allow shorter exposure times, producing a less blurred image.

While the Moon may look reasonably large to the unaided eye, it is only half a degree across. A 28mm camera lens used for everyday photography will produce an image of the Moon that is just a quarter of a millimeter in diameter on the film – far too tiny to show any detail when the photograph is printed or enlarged. By increasing the focal length of the telephoto lens, the size of the Moon's image projected onto the film increases. A tripod-mounted SLR and a telephoto lens with a focal length from 800 to 2,000mm will deliver the best telephoto lunar photographs. An 800mm lens will produce a lunar image over 7mm across on film, while an image larger than 18mm is produced using a 2,000mm telephoto lens. A larger telephoto lens will need to be attached to an equatorially driven mount, either directly or riding piggyback on a driven telescope.

An accessory called a teleconverter increases the focal length of a telephoto lens (see Table 2.2). A ×2 teleconverter will double the focal length of the lens, effectively doubling the range of available focal lengths in the photographer's armory. While an image of the Moon taken through, say, a 500mm telephoto lens in conjunction with a ×2 teleconverter will be the same size as one taken through a regular 1,000mm telephoto lens, the teleconverter image may be slightly dimmer and less well resolved, as a slight degree of optical degradation is introduced when the light passes through the extra lenses of the teleconverter. It is, however, far better to use a teleconverter to produce a larger lunar image than to enlarge an image produced by a smaller telephoto lens. However sharp the image may appear on the negative, enlarging it makes the image less distinct, and a point comes when the grain of the film also becomes visible.

Table 2.2. Telephoto photography and lunar image size on 35mm film

Focal length of lens (mm)	Image size on film (mm)	With ×2 teleconverter (mm)
28	0.25	0.51
50	0.45	0.91
100	0.91	1.82
135	1.23	2.45
200	1.82	3.64
300	2.73	5.45
500	4.55	**9.09**
800	**7.27**[a]	**14.55**
1,000	**9.09**	**18.18**
1,600	**14.55**	**29.10**
1,800	**16.36**	**32.72**
2,000	**18.18**	36.36

[a] Bold figures represent an optimum of lunar sizes on film, from a reasonably detailed image (over 7mm diameter) showing craters along the terminator to the best that can be taken using an undriven telephoto lens and fast film (up to 33mm diameter).

To calculate the size of the lunar image on film, simply divide the focal length of the lens (in millimetres) by 110.

A useful technique of eliminating the vibrations caused when the shutter is pressed is to use a dark card held in front of, but not touching, the telephoto lens. The cable release is pressed while the card is held in front of the lens, and a few seconds are allowed to elapse before the card is quickly withdrawn from the lens and then moved back again, say, a second later. The cable release is then pressed again to end the exposure. This may seem a hit-and-miss affair, but it is particularly effective when photographing the Moon at medium magnification with an equatorially mounted instrument and very low-speed film, such as 25 ISO. Excellent results can be obtained using Kodak Technical Pan 2415 (for black-and-white photographs) and Kodak Kodachrome 25 (for color slides). Both of these films are 25 ISO and produce sharp, contrasty images.

Prime Focus Photography

When a camera body (a camera minus its lens) is attached to a telescope (minus its eyepiece), the telescope acts in exactly the same fashion as a telephoto lens. Astrophotographers call this "prime focus photography," since the light falling onto the film is at the main focal point of the telescope's objective lens (or mirror). Prime focus photography delivers a much smaller image of the Moon on a 35mm frame of film than may be expected: a telescope with a focal length of under 2,000mm is capable of projecting the whole of the Moon's disc, half a degree in apparent diameter, onto a 35mm frame. Even a really big looking telescope – say, a 300mm f/6 Newtonian, with a focal length of 1,800mm – will not be able to fill the whole of a 35mm film frame with a prime focus image of the Moon. A Barlow lens inserted into the telescope's eyepiece holder will act in the same manner as a teleconverter used with a telephoto lens. Barlow lenses effectively increase the focal length of a telescope, and standard ones come in ×2 and ×3 varieties.

Although the Moon's image seen through the SLR viewfinder may seem awfully small, and details may be difficult to make out using short-focal-length telescopes, even with a Barlow lens, focusing the Moon at prime focus can be done using the camera's viewfinder and adjusting the telescope's own focuser. The process is usually quite forgiving, and by no means as exacting as focusing an image using eyepiece projection (see below). Prime focus is also extensively used with webcams and astronomical CCD cameras, but since CCD chips are so tiny in comparison to film, the magnification of the Moon that they deliver is much larger than on standard film (see below for CCD photography), and precise focusing is critical.

Prime focus exposure times must be gauged according to the telescope's focal length, the ISO of the film used and the phase of the Moon. As with all conventional lunar photography, it is advisable to "bracket" exposures – taking a number of images with a variety of exposure times. When beginning in astrophotography, it is also essential to make a written record of the precise exposures, film type and lunar phase so that you can identify an optimum combination that can be repeated in the future.

Eyepiece Projection

Close-up, high-magnification photographs of the Moon can be achieved by inserting an eyepiece into the telescope and then projecting the image into the camera, minus its lens. To do this, use widely available adapters that fit into standard-sized eyepiece holders (1.25-inch and 2-inch diameters) and the bodies of various makes of SLR. Orthoscopic and Plössl eyepieces deliver crisp images with flat fields of view. Eyepiece projection will deliver a far-higher-magnification image of the Moon than prime focus photography, the degree of magnification depending on the focal length of the telescope and the eyepiece and the distance of the eyepiece from the film plane. Shorter-focal-length eyepieces will deliver higher magnifications, and increasing the distance of the eyepiece from the film plane will also increase magnification. Focusing is achieved by looking directly at the magnified image of the Moon through the camera viewfinder and adjusting the telescope's focusing knob until a sharp image is seen. As with afocal photography, vignetting can appear around an image, depending on the type of eyepiece used and its distance from the film plane. If vignetting is present, it will be evident through the camera viewfinder.

Exposure Times

Formulas used for calculating lunar exposure times take into account the aperture and focal length of the lens/telescope, the camera f-stop used (if using a telephoto lens), the film's ISO rating and the Moon's phase. Because the Moon is a rough sphere reflecting sunlight to the observer on the Earth, the Moon's surface does not remain at a constant brightness, and it surprises many to learn that a half Moon is not half as bright as a full Moon – it is just one-ninth as bright. Exposure times therefore need to be altered according to the Moon's phase. The Moon's altitude above the horizon also affects its apparent brightness: a low Moon may be considerably dimmer than a high Moon because of and the presence of haze, cloud or atmospheric pollution. Given all of these variables, knowledge, experience and intuition all count a great deal when it comes to taking good photographs of the Moon.

Based on conventional lunar photography using a 100mm f/10 Maksutov as a telephoto lens, a ×2 Barlow lens and a standard SLR, giving a lunar image size of around 18mm on film, the following table may be of use in gauging the scale of exposure times and how it varies with ISO and lunar phase. Of course, these figures are not set in stone, and the photographer is advised to bracket photographs and make notes on which combination of telescope, film type and exposure times work best.

Film ISO	Narrow crescent	Broad crescent	Half Moon	Gibbous	Full Moon
25	1	1/2	1/4	1/8	1/15
50	1/2	1/4	1/18	1/15	1/30
100	1/4	1/8	1/15	1/30	1/60
200	1/8	1/15	1/30	1/60	1/125
400	1/15	1/30	1/60	1/125	1/250
800	1/30	1/60	1/125	1/250	1/500
1600	1/60	1/125	1/250	1/500	1/1000
3200	1/125	1/250	1/300	1/1000	1/2000

Digital Imaging

A CCD is a small flat chip, about the diameter of a matchhead in most commercial digicams, made up of an array of tiny light-sensitive elements called pixels. CCDs in low-end digicams may have an array of 640 × 480 pixels, while a more expensive digicam CCD may boast 2240 × 1680 pixels (around 4 megapixels). Light hitting each pixel is converted to an electrical signal, and the intensity of this signal directly corresponds to the brightness of the light that struck it. This information can be stored digitally in the camera's own memory or transferred to a PC, where it can be processed into an image. Digital images are infinitely easier to enhance and manipulate than conventional photographs in a photo lab darkroom.

Digicams, digital camcorders, webcams and dedicated astronomical CCD cameras have afforded amateur astronomers with quite modest equipment the opportunity to obtain wonderfully detailed images of the Moon. Indeed, just about anyone can take an acceptable lunar snapshot by pointing a digital camera through a small telescope. It's one thing to capture an image of the whole Moon, showing features along its terminator, but it requires considerable skill and expertise, both in the field and later at the computer, to produce a high-resolution, close-up image of the Moon capable of impressing a seasoned lunar observer. Digital imaging devices are not all identical. For example, afocal digicam imaging requires different techniques from those used imaging the Moon with a webcam and Barlow lens at prime focus.

Digital Cameras

Most low-end digicams have a fixed optical system, with non-removable lenses, and some may not even have an liquid crystal display (LCD) screen, so the principles involved in photographing the Moon through these cameras are just the same as with afocal photography using conventional fixed-lens film cameras (see above).

As a general rule, the digicam with the highest pixel rating will produce the clearest, highest-resolution views of the Moon. Most digicams can take images at a number of resolutions. A low-resolution image will take up less space in the camera's memory, and more images can be stored. Lunar photography requires the maximum resolution possible, so the setting should always be on high.

Most digicams provide instant results, their stored images being viewable on the camera's small LCD screen. The photographer can review each image individually to decide whether it's good enough to keep or whether to delete it and free the memory for a better image. Thin-film transistor (TFT) screens are usually on the small side, and the display is coarser than the captured image itself. To reduce time spent focusing the image, first focus the Moon in the eyepiece using your own eyes. When the digicam is fixed to the same eyepiece, the focus ought to be about right. Fine focusing is best achieved by viewing along the lunar terminator, where most relief detail is thrown up by the low elevation of the Sun and features appear at their sharpest. If the digicam has a zoom, the lunar terminator can be magnified, aiding focusing further. By hooking a digicam to a TV monitor or PC screen, the problems of focusing using the camera's inbuilt LCD screen are bypassed, as a larger image is far easier to focus.

Digicams are designed for everyday use, and their automatic settings may pose considerable problems when attempting to photograph the Moon, so it's essential to experiment with the digicam's settings to produce the best results. One of the issues to get to grips with is the digicam's exposure settings, as many afocal lunar images tend to be somewhat overexposed, the bright part of the Moon appearing washed out and lacking in any detail. The digicam's automatic exposure works best if there is a uniformly bright image across the entire field. A digicam may judge exposures perfectly well when the Moon is centered in the field of view or when taking close-up shots.

Color images of the Moon taken with digicams may show particularly vivid tones that aren't visible through the telescope eyepiece. While color can enhance the aesthetic quality of an image, it can also be undesirable. Computer processing can easily tone down any colors in an image. Capturing the Moon's colors in either an exaggerated or visually realistic fashion may produce a pleasing image, but the same amount of topographic detail can be recorded in black and white. If your camera has the ability to take black-and-white images, try it – the results may be noticeably sharper than those taken in color. Black-and-white images will take up less space in the camera's memory, too.

Using the digicam's zoom facility (if it has one) can eliminate the problems of image vignetting that tend to plague afocal photographs. Zooming adjusts the position of the camera's internal lenses – the magnified image becomes progressively dimmer, and vibrations in the telescope will show up more. When digital zoom comes into play at high magnifications, the quality of the image begins to degrade and the advantages of zooming are completely cancelled out. Optimum zoom is not maximum zoom; the amount of zoom used depends on the viewing conditions, the resolution of the CCD chip and the telescope, as well as the stability of the system and the accuracy of the telescope drive.

Camcorders

Camcorder footage of the Moon viewed in real time conveys a striking impression of actually looking through the telescope eyepiece. The viewer is using the same cerebral processing to make sense of the lunar landscape as in a real observation, as atmospheric shimmering distorts the view and the eye fixes upon individual objects to attempt to make out the fine detail. A high-magnification camcorder sweep along the Moon's terminator, taken under good viewing conditions and using a well-aligned driven telescope and slow-motion controls, enables the viewer to experience the grandeur of the Moon in a way that viewing still photographs cannot convey. It is possible to produce a wonderful tour of the Moon and its terminator, taking time to linger over features of interest and zoom in on them. Such footage makes a fantastic presentation at any astronomical society meeting, but think twice before showing all of your hard-won lunar camcorder footage to visitors and relatives, as they may not appreciate half an hour of wandering around the crater-crowded southern uplands as much as you do!

Camcorders have fixed lenses, and lunar footage must be obtained afocally through the telescope eyepiece. The same problems that affect afocal imaging using conventional film cameras and digicams apply to camcorders. Camcorders tend to be somewhat heavier than digicams, and it is essential that the camcorder be coupled to the telescope eyepiece as sturdily as possible. Some of the same equipment designed to hold digicams in place when taking afocal lunar images can be used to secure a lightweight camcorder.

Digital camcorders are the lightest and most versatile camcorders, and their images can be easily transferred to a computer for digital editing, employing the same techniques used with images obtained with a webcam (see below). Once downloaded onto a computer, individual frames from digital video footage can be sampled individually (at low resolution), stacked using special software to produce detailed, high-resolution images, or assembled into clips that can be transferred to a CD-ROM, DVD or videotape. The process can be time-consuming – the time spent running through the video footage and processing the images can amount to far longer than the time actually spent taking the footage. Digital video editing also consumes a great deal of a computer's resources, both in terms of memory and storage space; the faster a computer's CPUs and graphics card, the better. For basic editing of short video clips, at least 5 gigabytes are required on your computer's hard drive.

Webcams

Although they are designed mainly for use in the home to enable communication between individuals over the Internet, webcams can be used to capture high-resolution images of the Moon and planets. Priced at just a fraction of the cost of dedicated astronomical CCD cameras, webcams are extremely lightweight and versatile. Any commercial webcam hooked up to a computer and a telescope can be used to image the Moon and the brighter planets. Electrical signal noise greatly hampers the use of webcams to capture faint deep-sky objects like nebulae and galaxies, but the electronic circuitry within some makes of webcam have been modified by some amateurs to take longer exposures with reasonable success. The Moon is such a bright object that webcams can produce brilliant lunar images.

Although webcams may not have as sensitive a CCD chip as a more expensive astronomical CCD camera, their ability to record video clips made up of hundreds or thousands of individual images gives them a distinct advantage over a single-shot astronomical CCD. By taking a video sequence, the effects of poor viewing conditions can be overcome by selecting (either manually or automatically) the clearest images in the clip. These images can then be combined using stacking software to produce a highly detailed image. This may show as much detail as a visual view through the eyepiece using the same instrument.

It is possible to photograph the Moon afocally with webcams, but they are usually used at the telescope's prime focus, minus the webcam's original lens. A number of webcams have easily removable lenses, and commercially available adapters can be screwed in their place, permitting easy attachment to a telescope. Some webcams, however, require disassembly to remove the lens, and the adapter needs to be home-made. CCDs are sensitive to infrared light, and the lens assembly may contain an infrared-blocking filter; without the filter, a really clean focus through a refractor is not possible, since infrared is focused differently from visible light. Infrared-blocking filters are, however, available for fitting between the telescope and webcam, allowing only visible-light wavelengths to pass through for a sharp focus.

Like the CCD chips in most other digital devices, webcam CCDs are very small – 3.6 × 2.7 mm in a Philips ToUcam Pro. Used at the telescope's prime focus, the image of the Moon that webcams produce is at quite a substantial magnification, and the entire Moon can only be captured in the field of a very short-focal-length telescope, in the region of 250mm. Because of the high magnifications produced by a webcam at prime focus on an average amateur telescope, a well polar-aligned driven equatorial telescope with electric slow motion controls is essential in order to capture a relatively static video clip lasting 10 or 20 seconds. If the image drifts too much during the clip, the software used to process the video clip may not be able to produce a good alignment.

In order to calculate the focal ratio of the light path that is optimally suited to capture all the fine lunar detail that a particular telescope and webcam can to resolve, multiply the pixel size of the CCD chip (in microns) by 3.55. The pixel size of a webcam is usually given in the technical information that accompanies it or on the manufacturer's website. For example, in a Philip's ToUcam Pro the pixel size is 5.6 microns; multiplying this by 3.55 gives an optimum focal ratio of 19.88. For an f/6 telescope, the nearest focal ratio to this can be obtained by inserting the webcam into a ×3 Barlow lens, giving a focal ratio of f/18. In an f/10 telescope it would be best to use a ×2 Barlow lens, giving a slightly longer focal ratio of f/20. Note that this is an optimum formula based on excellent viewing conditions – poor visibility and high magnifications will just produce larger blurred images of the Moon. In mediocre viewing conditions, it is best to multiply the pixel size by 2 to arrive at a more suitable focal ratio. It would be better to use a ×2 Barlow than a ×3 one in an f/6 telescope to produce the best images in less than ideal viewing conditions, and no Barlow at all in an f/10 telescope.

Using basic equipment alone, focusing a webcam can prove to be time-consuming. To achieve a rough focus, it is best to set up during the daytime and focus on a distant terrestrial object using the telescope and webcam, viewing the computer monitor and adjusting the focus manually. If your telescope is some distance from the monitor, this may require a number of trips to and from the telescope and computer. A laptop in the field, near the telescope, would save a great deal of time, both during this initial process and during lunar imaging itself. Once the terrestrial object has been focused, lock the focus or mark the focusing barrel with a chinagraph pencil.

During the imaging session, the Moon is centered in the field of vision using the telescope's finder (which must be accurately aligned), and the Moon will appear on the computer screen, probably still requiring further focusing. It is best to focus on the lunar terminator, where features are most sharply defined. When the telescope's focus is adjusted manually, care must be taken not to nudge the instrument too hard, as the Moon may disappear altogether out of the small field of vision. Patient trial and error will eventually produce a reasonably sharp focus; once achieved, lock the focuser and mark the focusing tube's position so that an approximately sharp focus can be found quickly during subsequent imaging sessions.

Achieving the best focus makes the difference between a good lunar image and a great one, and a fraction of a millimeter can make the difference between a good focus and a tack-sharp one. Focusing by hand is exceedingly time-consuming, and a perfect focus is more likely to be found by chance than by trial and error. Electric focusers enable the focus to be adjusted remotely from the telescope, and they are considered an essential accessory to the lunar imager, not a luxury. Electric focusers save a lot of time, making a great difference to your enjoyment of imaging, but, more important they offer infinitely more control over fine focusing. A webcam attached to a computer through a high-speed USB port will deliver a rapid refresh rate of the image, enabling fine focusing in real time.

Video sequences of the Moon can be captured using the software supplied with the webcam. It is necessary to override most of the software's automatic controls: contrast, gain and exposure controls require adjusting to deliver an acceptable image. Many imagers prefer to use the black-and-white recording mode, which cuts down on signal noise, takes up less hard drive space and eliminates any false color that may be produced electronically or optically.

Most webcams can record image sequences using frame rates of between 5 and 60 frames per second (fps). A 10-second video clip made at 5 fps will be composed of 50 individual exposures and may take up around 35 MB of computer memory. At 60 fps there will be 600 exposures, and the amount of space taken up on the hard drive will be proportionately greater. Using the webcam's highest resolution (in most cases, an image size of 640×480), a frame rate of between 5 and 10 fps and a video clip of 5–10 seconds is optimum. If 5–10 of these clips, centered on the same area, are secured in quick succession, the imager will have between 125 and 1,000 individual images work from. The sheer number of images provided by webcams is their greatest strength. A single-shot dedicated astronomical CCD camera costing perhaps ten times as much as a webcam only takes one image at a time. An image produced by an astronomical CCD may have far less signal noise and a higher number of pixels than one taken with a webcam, but in mediocre viewing conditions, the chances that the image was taken at the precisely right moment are small. Webcams can be used even in poor viewing conditions, as a number of clearly resolved frames will be available to use in an extended video sequence. Video sequences are usually captured as Audio Video Interleave (AVI) files.

Astronomical image-editing software is used to analyze the video sequence, and there are a number of very good free imaging programs available. Some programs are able to work directly from the AVI, and much of the process can be set up to be automatic – the software itself selects which frames are the sharpest, and these are then automatically aligned, stacked and sharpened to produce the final image. If more control is required, it is possible to individually select the images in the sequence to be used; since this may require up to a 1,000 images to be visually examined, one after another, it can be a laborious process, but it can produce sharper images than those derived automatically. Images can be further processed in image manipulation software to remove unwanted artifacts, to sharpen and enhance an image's tonal range and contrast and to bring out detail. Unsharp masking is one of the most widely used tools in astronomical imaging: almost magically, a blurred image can be brought into sharper focus. Too much image processing and unsharp masking may produce spurious artifacts in the image's texture and a progressive loss of tonal detail.

Chapter 5

Viewing the Moon with the Unaided Eye

Features Visible on the Moon

To the unaided eye, the illuminated face of the Moon appears as a patchwork of dark and light areas. When it is high in a clear sky, the glare of a full Moon can be too bright for much fine detail to be discerned with the unaided eye, but if the Moon is observed when it is close to the horizon, veiled by thin cloud or viewed through sunglasses, a surprising amount of detail may be discerned. The Moon's seas cover a third of the near side and appear as dark patches to the naked eye, and a range of marial tones is evident. The brighter areas surrounding the maria are mountainous cratered regions and the light-colored ray systems of a few impact craters. Some of these bright ray systems can be discerned without optical aid.

Lunar features may be used as a test of visual acuity. Starting with the easiest, becoming progressively more difficult, suggested targets are:

1. Bright region around Copernicus
2. Mare Nectaris
3. Mare Humorum
4. Bright region around Kepler
5. Gassendi region
6. Plinius region
7. Mare Vaporum
8. Lubiniezky region
9. Sinus Medii
10. Faintly shaded area near Sacrobosco
11. Dark spot at foot of Mt Huygens
12. Riphean Mountains

Those with good eyesight will certainly be able to discern the bright area formed by the rays around crater Copernicus, and the dark lava patch of Mare Vaporum can be made out with little trouble. Last on the list, the Riphean Mountains represent a challenge for those with superb eyesight since the light gray cluster of peaks only subtends an apparent angle of around 2 arcminutes.

Those with good vision can comfortably discern the bright ray craters Copernicus, Kepler, Tycho, Langrenus and Aristarchus at full Moon, but the Moon's terminator is smooth to most people's eyes. Irregularities along the Moon's terminator can sometimes be seen by those with very good eyesight; the most pro-

Fig. 5.1. A test of visual acuity – twelve features visible on the full Moon with the unaided eye (numbered easiest to hardest). The following are in the text of the book: **1** Bright region around Copernicus; **2** Mare Nectaris; **3** Mare Humorum; **4** Bright region around Kepler; **5** Gassendi region; **6** Plinius region; **7** Mare Vaporum; **8** Lubiniezky region; **9** Sinus Medii; **10** Faintly shaded area near Sacrobosco; **11** Dark spot at foot of Mt Huygens; **12** Riphean Mountains. Those with good eyesight will certainly be able to discern the bright area formed by the rays around crater Copernicus, and the dark lava patch of Mare Vaporum can be made out with little trouble. Last on the list, the Riphean Mountains represent a challenge for those with superb eyesight since the light grey cluster of peaks only subtends an apparent angle of around 2 arcminutes.

minent of these becomes apparent when Sinus Iridum has just emerged into the morning sunshine when the Moon is around 10 days old, with the brilliant curve of the bordering Montes Jura projecting into the darkness beyond. The big craters Ptolemaeus near the center of the Moon's disc, and Clavius, in the southern uplands, also make dents on the terminator every two weeks, though excellent eyesight is necessary to make them out.

With the naked eye, it is quite possible to monitor some effects of the Moon's libration, which noticeably affects the apparent proximity to the limb of the dark oval-shaped Mare Crisium in the east and the narrow dusky streak of Mare Frigoris in the north. Those with average eyesight will find that both Mare Crisium and Mare Frigoris, when they are closest to the lunar limb, are something of a challenge to see clearly, but both are easy objects to spot during favorable librations.

Earthshine

Earthshine is the faint illumination of the Moon's dark side caused by sunlight being reflected onto the Moon by the Earth. The phenomenon is most obvious when the Moon is a thin, waxing crescent in a reasonably dark sky, a sight sometimes referred to as "the old Moon in the young Moon's arms." Viewed from the Moon's surface at this time, the Earth would appear as a bright waning gibbous sphere lighting the lunar landscape with up to 60% more brilliance than our own full Moon.

With the naked eye, earthshine may be followed for several days until around first quarter phase and picked-up again in the last quarter of the lunar month. The brightness of earthshine actually varies according to Earth's phase, geography and global weather conditions continents are more reflective than oceans, and a cloudy globe is highly reflective. Anyone who regularly observes earthshine will soon come to notice that there are times when the phenomenon seems especially prominent.

Perception Is Not Always Reality

Although the human eye is in many ways like a digital camera, in that a small lens system projects photons onto a matrix of sensitive light receptors, what the observer perceives is not always reality. The raw image projected onto the retina is processed by the brain, sampled and analyzed to produce an image that makes the most sense to the observer. Optical illusions can fool the most experienced observer, and some illusions are so powerful that they are difficult not to have, even though the observer may be fully aware of their nature.

Moon Illusion

Of all the visual tricks played upon the observer, Moon illusion is the best known. When the full or near-full Moon is seen close to the horizon, it may appear to assume unusually large proportions. Some people call such seemingly huge Moons "as big as a plate." In reality, the Moon's apparent diameter actually becomes increasingly *smaller* the closer it is to the horizon, because the observer is viewing the Moon from a spherical Earth. When the Moon is high overhead it can be more than 6,000 km closer to the observer than when it appeared rising above the eastern horizon on the same night. This difference in the Moon's apparent diameter is far too subtle to be detected with the naked eye, as it amounts to around 1 minute of arc at maximum. The Moon illusion occurs mainly because we perceive the sky as being shaped like the interior of a flattened dome; this is reinforced by the perspective effect of clouds in the sky. Celestial objects seem to be attached to the interior of the celestial dome. When the near-horizon Moon is viewed, we imagine that it is far away and must therefore be a large object to subtend an angle of around half a degree. When the Moon is above us, we subconsciously imagine it to be closer to us and much smaller in size. Exactly the same illusion occurs in our perception of constellations. Castor and Pollux, for example, are always 4.5° apart, but they can appear to be more widely separated when they are near the horizon than when they are high in the sky.

Lunar Cross-Staff

By observing the Moon with a simple naked eye measuring device called a cross-staff, it can be determined that that the Moon remains more-or-less the same size whether it is riding high in the sky or hovering over the distant horizon. Cross-staffs and assorted naked eye devices were used extensively in the pretelescopic era, and they are easy to construct. To a thin, straight length of wood about a meter-long, attach a small slider that can be moved up and down the main staff. The slider should be easy to move, though not so loose as to slide down the staff of its own accord when pointed upwards. At the leading edge of the slider, fasten two straight pins or small nails side by side, 8 mm apart. Along the upper half of the cross-staff, draw a line 30 cm long and mark it at 1 cm intervals. At the other end of the staff fasten a small squint holel a used cotton reel or piece of card with a small hole punched out will do. The instrument is now ready to be turned to the Moon.

With the staff resting on a solid surface (say, a garden fence or a camera tripod) the Moon is sighted through the squint hole and aligned between the pins on the slider. If the Moon is not quite full, rotate the staff slightly so that the widest illuminated breadth of the Moon is lined-up between the pins, or with the tips of the crescent if the Moon is only narrowly illuminated. Make a note of the slider's position on the 30 cm scale marked on the staff. It will be seen that the Moon's apparent diameter always remains around half a degree regardless of its location in the sky.

Throughout the month the Moon's distance ranges from 356,400 km when nearest the Earth (perigee) to 406,700 km when furthest away (apogee). This difference of 50,300 km means that the Moon's apparent diameter at perigee is 33′ 29″, compared with 29′ 23″ at apogee – the perigee Moon presents an area nearly 30% larger than at apogee. Such a small variation in apparent diameter, even in an object like the Moon, only around half a degree wide, can be detected with the lunar cross-staff. At perigee, when the Moon is at its largest apparent diameter, the slider's leading edge will be about 83 cm from the squint hole. At apogee the slider will be around 94 cm from the squint hole. If the Moon's diameter is measured in this manner over a period of a year, then it becomes clear that the period from one perigee to the next; called the "anomalistic month," is shorter than the synodic month (new Moon to new Moon) by a couple of days.

Tracking the Moon

Some careful thought is needed when considering the Moon's visibility and its height above the horizon at various times of the year. The Moon moves from west to east, each hour traveling an angular distance a little more than its own diameter. In 29.5 days it goes through a complete set of phases, from new Moon, through full Moon and back to new Moon again; this period is called a synodic month, or lunation. Because the plane of the Moon's orbit lies close to the ecliptic (inclined just 5°

to it), the Moon's monthly path among the constellations of the zodiac is similar to that followed by the Sun over the course of a whole year.

The full Moon is, of course, always opposite to the Sun. At winter solstice, when the Sun is at its lowest point above the horizon, the mid-winter full Moon rides high in the early morning skies. At summer solstice, the Sun reaches its highest point above the horizon, whereas the mid-summer full Moon just manages to pull itself clear of the southern horizon for a few hours (as seen from northern temperate latitudes such as the U.K., the northern U.S. and southern Canada). Around the time of full Moon in autumn, the ecliptic (and the Moon's orbital plane) is at its shallowest angle to the eastern horizon, causing the Moon to rise only a few minutes later each evening. The 'Harvest Moon' is the full Moon nearest to the date of autumnal equinox on 23 September. In spring the ecliptic is ar a steep angle to the eastern horizon, which means that, around the time of full Moon, its consecutive evening rising times are widely separated; sometimes by as much as an hour and a half.

Color Perception in Moonlight

Though the full Moon appears big and bright, sunlight is half a million times brighter. A terrestrial landscape bathed in moonlight has a ghostly monochrome appearance because the scene is not bright enough to trigger all of the color receptors in the human eye. To test this, prepare 10 pieces of different-colored paper – say, white, gray, pink, light brown, pale green, lilac, sky blue, bright red, bright green, and yellow – and label them on the back. On a clear night, select a dark spot outside that is illuminated only by the light of the Moon and attempt to identify each color (after shuffling the paper). Red may turn out to be the only color that can be identified with certainty, and only then if the moonlight is bright enough.

Those living near sodium streetlighting will be able to perform another experiment, this time illustrating a psychological aspect of human color perception. When the Moon is high in the sky and near full, stand close to an orange streetlamp, and face the Moon. Your shadow, cast on the ground by the lamp, will seem to be indisputably blue, though in reality your shadow is being faintly illuminated with moonlight which is white, not blue. This odd effect comes about because our brains are wired to acknowledge that the brightest source of illumination is white light – in nature, the Sun and the Moon. Therefore, we perceive our sodium orange surroundings as being illuminated with white light, and the blueness of the shadow is simply caused by color contrast perception – an effect obvious in double stars like Albireo (Beta Cygni), whose brighter yellow component much enhances the blueness of its companion.

Atmospheric Effects

Lunar rainbows form under the same circumstances as those of the Sun; both are colored arcs of 42° radius situated diametrically opposite the light source. Lunar rainbows, however, have a maximum brilliance of a mere 1/500,000 of their daytime counterparts and, as such, they are rare, dim and somewhat hypochromic.

The white corona often seen in the Moon's immediate vicinity is caused by moonlight's reflection and diffraction amid water droplets in the lower clouds. The lunar corona is sometimes encircled by one or two and, on rare occasions, three rainbow-colored halos caused by moonlight's diffraction by countless water droplets or ice crystals in the upper atmosphere. Lunar halos can be particularly vivid when moonlight passes through a thin homogenous cloud layer composed of very minute water droplets. Sometimes the lunar halo may be host to parselene (also known as "mock Moons" or "Moon dogs" diffuse images of the Moon located 22° to either side of it, caused by the moonlight's refraction by ice crystals in the upper atmosphere.

Chapter 6

Lunar Showcase: A Binocular Tour of the Moon's Trophy Room

Binoculars reveal far more of the Moon than does the naked eye. Through them, a distant disc with a few dark and light patches is transformed into a spherical globe full of detail, splashed with dark areas and pockmarked with craters. Although amateur astronomers tend to underrate binoculars, they have a number of advantages even over telescopes. Binoculars (with the exception of giant ones) are eminently portable and are capable of withstanding occasional knocks. Despite delivering a field of view that may be a dozen or more times the diameter of the Moon, a steadily held pair of binoculars will reveal a considerable amount of detail, both in the brighter parts of the lunar disc and along the terminator. Viewing the Moon with two eyes is a more pleasurable experience than using just one: even though the true 3-dimensional nature of the Moon is not possible to discern because the parallax effect is far too small, two eyes get the impression of a globe floating in space.

Simple opera glasses (also known as Galilean binoculars) deliver very low magnification and a narrow field of vision, and a degree of false color becomes apparent when bright objects like the Moon are viewed. Opera glasses will show all the Moon's maria, some of the major mountain ranges and dozens of large craters, including several surrounded by bright ray systems. They are useful for quick peeks at the Moon in order to plan telescopic observing sessions.

A pair of good-quality binoculars reveal a considerable amount of lunar detail. A binocular view of the crescent Moon, its dark side adorned with blue earthshine, is utterly beautiful. Binoculars can be used to observe the brighter lunar occultations of stars and planets. Lunar eclipses are better viewed through binoculars than through a telescope: both eyes can allow the subtle colors in the umbra to be readily discerned, and the totally eclipsed Moon hovering against a starry background often appears stunningly 3-dimensional.

Examples of just about every type of lunar feature can easily be discerned through a well-mounted pair of binoculars or a small refractor. A 50mm instrument is capable of resolving features as small as 10 km across. For binocular studies of the Moon, the higher the magnification, the better; with the exception of zoom binoculars and specialist giant binoculars, most binoculars give magnifications of between 7 and 15×. The highest usable magnification on a 50mm refractor is around 100x, but it is difficult to keep the Moon centered in the field of

Fig. 6.1. General map of the Moon, showing all the main features visible through binoculars. Credit: Peter Grego

an undriven altazimuth instrument at such a high magnification. It is better to use a magnification that gives a satisfactorily detailed view of the Moon comfortably and easily viewed.

The following is a selection of some of the most prominent features of each type of lunar feature, suitable for observation through binoculars or a small telescope, with particular attention paid to the most prominent feature in each class. The list below is not meant to be exhaustive, and more detailed information on each object is contained in Chapter 7.

Maria: Vast Seas of Solidified Lava

One-third of the Moon's near side lies buried beneath maria (Latin: seas) – dark lava flows that fill the interiors of a number of huge asteroidal impact basins. Many of the maria are roughly circular in outline, while others are somewhat irregular, the lavas having extended into outlying plains and adjoining craters.

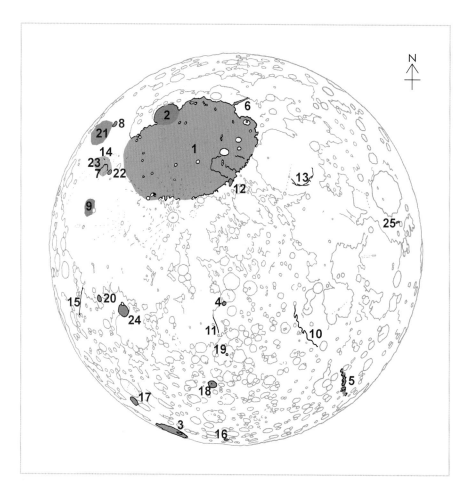

Fig. 6.2. Some of the largest and best, record-breaking lunar features.
1 Largest circular mare: Mare Imbrium (3,554,000 sq km). **2** Largest bay: Sinus Iridum (167,000 sq km). **3** Largest crater: Bailly (303 km). **4** Largest central peak: in Alpetragius (2,910 m). **5** Largest valley: Vallis Rheita (500 km long, 50 km wide max.). **6** Largest graben valley: Vallis Alpes (180 km long, 18 km wide max.). **7** Largest sinuous rille: Vallis Schröteri (160 km long, 10 km wide max). **8** Largest dome: Mons Rümker (70 km wide). **9** Largest field of domes: Marius hills (area 30,000 sq km). **10** Largest normal fault: Rupes Altai (480 km long). **11** Neatest normal fault: Rupes Recta (110 km long). **12** Highest mountain: Mons Huygens (5,400 m). **13** Largest wrinkle ridge: Dorsa Lister (300 km long). **14** Smallest wrinkle ridge: Dorsum Niggli (50 km long). **15** Longest rille: Rima Sirsalis (400 km long). **16** Deepest crater: Newton (8,839 m). **17** Most lava-filled crater: Wargentin (84 km). **18** Crater with most prominent rays: Tycho (85 km, rays up to 1,300 km long). **19** Brightest area: 'Cassini's Bright Spot' (Deslandres HA). **20** Darkest area: Billy (46 km). **21** Blandest area: junction of Sinus Roris and Oceanus Procellarum. **22** Most TLP-prone area: Aristarchus (40 km). **23** Most noticeably colored area: "Wood's Spot" near Aristarchus (orange). **24** Most frequently observed crater: Gassendi (110 km). **25** Smallest lake: Lacus Perseverantiae (70 km across). Credit: Peter Grego

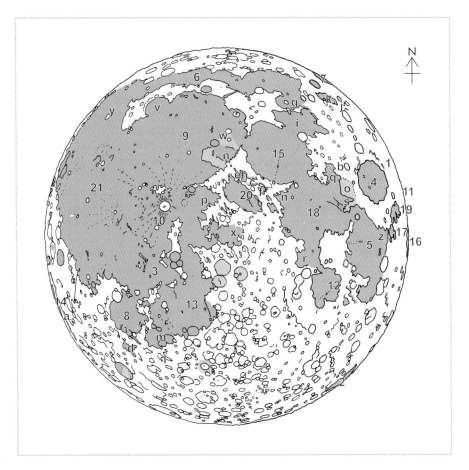

Fig. 6.3. Map showing the Moon's near-side maria, marshes, lakes and bays, all visible through binoculars.

Key: **Maria: 1** Mare Anguis **2** Mare Australe **3** Mare Cognitum **4** Mare Crisium **5** Mare Fecunditatis **6** Mare Frigoris **7** Mare Humboldtianum **8** Mare Humorum **9** Mare Imbrium **10** Mare Insularum **11** Mare Marginis **12** Mare Nectaris **13** Mare Nubium **14** Mare Orientale **15** Mare Serenitatis **16** Mare Smythii **17** Mare Spumans **18** Mare Tranquillitatis **19** Mare Undarum **20** Mare Vaporum **21** Oceans Procellarum. **Lacus/Palus/Sinus – a** Lacus Autumni **b** Lacus Bonitatis **c** Lacus Doloris **d** Lacus Excellentiae **e** Lacus Felicitatis **f** Lacus Lenitatis **g** Lacus Mortis **h** Lacus Odii **i** Lacus Somniorum **j** Lacus Spei **k** Lacus Temporis **l** Lacus Timoris **m** Palus Epidemiarum **n** Palus Putredinis **o** Palus Somni **p** Sinus Aestuum **q** Sinus Amoris **r** Sinus Asperitatis **s** Sinus Concordiae **t** Sinus Fidei **u** Sinus Honoris **v** Sinus Iridum **w** Sinus Lunicus **x** Sinus Medii **y** Sinus Roris **z** Sinus Successus. Credit: Peter Grego

Oceanus Procellarum (Ocean of Storms). An elongated marial expanse occupying much of the far western lunar hemisphere, sharply defined in the west. Narrows into Sinus Roris (Bay of Dew) in the north, and its eastern reaches blend into Mare Imbrium. Sunrise: 10–13d. Sunset: 25–28d.

Mare Imbrium (Sea of Rains). The largest circular mare, with large, well-defined mountain borders for 300°. Blends with Oceanus Procellarum in the west. Northern border dented by the semicircular Sinus Iridum (Bay of Rainbows). Sunrise: 6–10d. Sunset: 21–25d.

Mare Serenitatis (Sea of Serenity). Circular, with well-defined mountain borders, links with Mare Tranquillitatis in the south and Lacus Somniorum in the northeast. Mare borders display interesting sharply defined tonal variations. Location of the Moon's most prominent wrinkle ridge (see below). Sunrise: 4–6d. Sunset: 19–21d.

Mare Tranquillitatis (Sea of Tranquillity). An irregular mare, somewhat rectangular in outline. Joined to Mare Serenitatis in the north, Mare Fecunditatis in the east and Sinus Asperitatis (Bay of Asperity) in the south. Location of the fascinating wrinkle feature Lamont (see below). Sunrise: 3–5d. Sunset: 18–20d

Mare Nubium (Sea of Clouds). A large, irregular mare in the southern central part of the Moon. Sunrise: 7–9d. Sunset: 22–24d.

Mare Fecunditatis (Sea of Fertility). An elongated mare with irregular borders occupying a large portion of the Moon's eastern near limb region, south of Mare Crisium. Its northwestern reaches join with Mare Tranquillitatis. Sunrise: 2–4d. Sunset: 16–18d.

Mare Crisium (Sea of Crises). Prominent circular mare near the northeastern limb, surrounded by a continuous mountain border. Looks like a very large crater when bisected by the terminator. Sunrise: 2–3d. Sunset: 16--17d.

Mare Humorum (Sea of Humors). Dark, circular sea in the southwestern quadrant of the Moon, its northeastern mountain border broken by the southern reaches of Oceanus Procellarum. Sunrise: 9–10d. Sunset: 25–26d.

Mare Nectaris (Sea of Nectar). A small, well-defined circular sea, its southern shoreline dented by the large bay crater Fracastorius (124 km). Mare Nectaris joins Sinus Asperitatis in the north. Sunrise: 4–5d. Sunset: 18–19d.

Mare Frigoris (Sea of Cold). A dusky tract stretching across a great swath of the Moon's northern latitudes, from Sinus Roris in the west to Lacus Mortis (Lake of Death) in the east. Sunrise: 4–10d. Sunset: 19–25d.

Mare Vaporum (Sea of Vapors). A small, irregular dark mare that occupies the highlands southwest of Mare Serenitatis. It links with the plains of Sinus Medii (Central Bay) in the south. Sunrise: 6–7d. Sunset: 21–22d.

Tour of Mare Imbrium: A Small Telescope Trek

Mare Imbrium is the Moon's largest circular sea, a vast expanse of lava occupying an asteroidal impact basin that was blasted out of the Moon's crust around 3.8 billion years ago. Mare Imbrium is 1,160 km across, has a circumference of 3,900 km and a surface area of 830,000 sq km.

Mare Imbrium is surrounded for 300° of its circumference by the most spectacular series of mountain ranges on the Moon's near side. Beginning in the northwest and progressing clockwise around Mare Imbrium, two massive mountain blocks, Mons Gruithuisen Gamma (base diameter 20km) and Mons Gruithuisen Delta (base diameter 25km) mark the beginning of Imbrium's northern mountain border. North of these brooding sentinels, cratered, hilly terrain rises from the flat expanses of Oceanus Procellarum (Ocean of Storms) to the west and blends into Montes Jura, a magnificent arc of peaks. The Jura mountain arc forms the northern border of the magnificent bay of Sinus Iridum (Bay of Rainbows), a 260-km-diameter asteroidal impact basin created after Mare Imbrium, which has been infilled with Imbrium lava flows. Observed through a small telescope when the Moon is around 11 days old, the southwestern portion of the Montes Jura takes on the appearance of a graceful art deco figurine, a simulacrum known as the "Moon Maiden." Her finely detailed facial profile is **Promontorium Heraclides** (Cape Heraclides) and her hair is the hilly ground further west. The rest of her body, composed of the Montes Jura, curves away at a graceful angle. Bear in mind that she can be seen with binoculars from the northern hemisphere, but appears upside down.

Undulating mountains continue north of Mare Imbrium and broaden to engulf the large, dark-floored **Plato**, a striking, flat-floored crater 100 km across. To its east, the broad mountain range of **Montes Alpes** (the lunar Alps) are sliced through by **Vallis Alpes** (Alpine Valley), a lunar rift valley 130 km long and 18 km wide in places (see below).

In Mare Imbrium, south of Plato, several mountain ranges and peaks are entirely surrounded by mare material. Because they are prominent and lie on flat ground, these features and the shadows cast by them are well placed for viewing through small instruments. **Montes Recti** (Straight Range) is a line of peaks 90 km long rising to 1,800 m – dimensions slightly smaller than those of the Black Hills of Dakota. East of Montes Recti are **Montes Teneriffe**, a cluster of mountains that rise to 2,400 m in places. The mountains on the terrestrial island of Tenerife in the Atlantic Ocean rise higher (up to 3,700 m) and are volcanic in origin. East of Montes Teneriffe is the prominent **Mons Pico**, an isolated mountain 2,400 m high, named in honor of the highest mountain on Tenerife. In northeastern Mare Imbrium, the 2,250 m high **Mons Piton** (also named after a peak on Tenerife) stands proudly above its surroundings. All of these mountain features are thought to predate the lava flows of Mare Imbrium, and represent the remnants of the Imbrium basin's inner ring. East of Piton, at the eastern end of the lunar Alps, lies the crater **Cassini** with its unusual extended outer flange and dark-flooded floor punctuated by two large craters. West of Cassini, **Montes Caucasus** is a substantial mountain range 550 km long, marking the boundary between northwestern Mare Serenitatis and eastern Mare Imbrium.

A beautiful quartet of features resides in the western part of Mare Imbrium; the flat-floored crater **Archimedes** (83km); the prominent impact crater **Aristillus** (55km); with its central peaks, terraced walls and ray system, the smaller crater **Autolycus** (39km), and **Montes Spitzbergen**. Montes Spitzbergen is 60 km long, with peaks up to 1,500 m high, named because of their resemblance to the terrestrial Spitzbergen Islands (Svalbard).

Montes Apenninus (Apennine Mountains) form the southeastern border of Mare Imbrium. Some individual peaks within this vast range rise above 5,000 m. Much of the Apennine range is striated in a pattern radial to Mare Imbrium, somewhat resembling the effects of glaciation. Of course, large glacial ice sheets have never heaved their way across the Moon's surface; the pattern around Mare Imbrium and other large impact basins was caused by secondary impact sculpting, combined with faulting. The eastern end of the Apennines is marked by the 60-km crater **Eratosthenes**, after which there is a gap in the mountain chain, but picked up 90 km further on by **Montes Carpatus** (Carpathian Mountains), which mark Mare Imbrium's southern border. The Carpathians are over 400 km long and are overlain by thick deposits of bright material ejected by the Copernicus impact some 900 million years ago (see below).

In addition to the splendid mountain ranges found in and around the borders of Mare Imbrium, there are numerous other prominent mountain ranges that can be

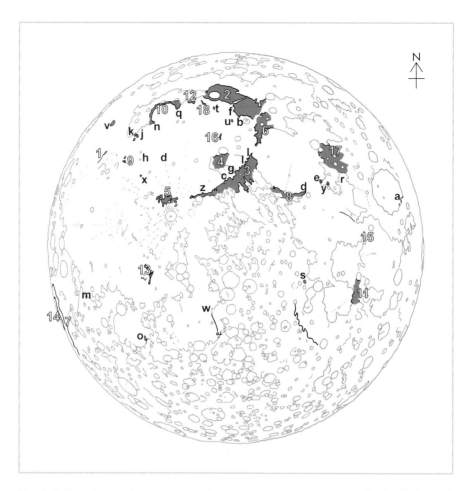

Fig. 6.4. Map showing the Moon's named mountain ranges, promontoria and individual named peaks.

Key: **Montes – 1** Agricola **2** Alpes **3** Apenninus **4** Archimedes **5** Carpatus **6** Caucasus **7** Cordillera **8** Haemus **9** Harbinger **10** Jura **11** Pyrenaeus **12** Recti **13** Riphaeus **14** Rook **15** Secchi **16** Spitzbergen **17** Taurus **18** Teneriffe. **Mons & Promontoria – a** Prom Agarum **b** Prom Agassiz/Deville **c** Mons Ampere/Huygens **d** Prom Archerusia **e** Mons Argaeus **f** Mons Blanc **g** Mons Bradley **h** Mons Delisle **i** Prom Fresnel **j** Mons Gruithuisen Delta **k** Mons Gruithuisen Gamma **l** Mons Hadley/Hadley Delta **m** Mons Hansteen **n** Prom Heraclides **o** Prom Kelvin **p** Mons La Hire **q** Prom Laplace **r** Mons Maraldi **s** Mons Penck **t** Mons Pico **u** Mons Piton **v** Mons Rümker **w** Prom Taenarium **x** Mons Vinogradov **y** Mons Vitruvius **z** Mons Wolff. Credit: Peter Grego

viewed through binoculars and small telescopes. Some of these were created as a result of basin-forming asteroidal impacts, while others are the remnants of crater rims and highlands that have been surrounded by lava flows.

Montes Caucasus. A prominent wedge of mountains containing some spectacular individual peaks. Montes Caucasus runs 500 km south from the large craters **Aristoteles** (87km) and **Eudoxus** (67km), marking part of the northwestern border of Mare Serenitatis. An easy object to see through binoculars, and very impressive when illuminated by a morning Sun. Sunrise: 7d. Sunset: 21d.

Montes Taurus. Under low illumination, a prominent highland region adjoining Mare Serenitatis's southeastern border, but appearing gray and muted under a high Sun. Sunrise: 4d. Sunset: 19d.

Montes Riphaeus. A compact, branching group of mountains 195 km long and shaped like the footprint of a great crested grebe, complete with webs. The range separates northwestern **Mare Cognitum** (the Known Sea) from Oceanus Procellarum. Their mare location makes them an easy group to discern even under a high Sun. Sunrise: 10d. Sunset: 24d.

Wrinkle Ridges

Here and there the maria are crossed by low wrinkle ridges (properly termed "dorsa"). Although these ridges are just a few tens of meters in height, they are visible when illuminated at a very oblique angle just after local sunrise or before sunset, and some of the larger wrinkle ridges can be discerned through binoculars and small telescopes. Most of the dorsa are compression features, formed when the mare lavas subsided, reducing the marial surface area and crumpling mare material in a gradual manner. Mare Crisium, Mare Humorum, Mare Imbrium, Mare Nubium, Mare Fecunditatis and Mare Tranquillitatis all display prominent systems of dorsa, most of which run parallel to the mare borders.

Dorsa Smirnov – The Serpentine Ridge. Perhaps the most well-known wrinkle ridge is **Dorsa Smirnov** (often termed the "Serpentine Ridge"), a large system of wid, braided ridges that run parallel to the eastern border of Mare Serenitatis. It is 200 km long and 15 km wide in places. Dorsa Smirnov can easily be glimpsed through binoculars or a small telescope when it is near the terminator. No trace of it can be seen through a telescope at high illuminations. Sunrise: 8d. Sunset: 20d.

Lamont – Unique Wrinkled Formation. A network of wrinkle ridges 300 km from north to south converges on **Lamont**, a unique and fascinating feature in southwestern Mare Tranquillitatis. The main ring of Lamont is 75 km in diameter and composed of a series of low dorsa, giving it the appearance of a bullethole in a sheet of toughened glass. Lamont is only visible at a low morning or evening illumination through a small telescope, though it is not an easy object. It cannot be discerned at all when illuminated by a high Sun. Sunrise: 5d. Sunset: 20d.

Domes

Low, rounded hills known as domes occur singly or in small clusters in parts of the maria. Like wrinkle ridges, domes do not rise steeply from the landscape and can only be seen when near the terminator, illuminated by a low Sun. Most domes are volcanic features built up by successive eruptions of lava that may have had a viscosity equivalent to that which formed (and is still forming) the Hawaiian shield volcanoes on Earth. It is possible, though, that some domes are uplift features like the Adirondack Mountains in New York State, where deep intrusions of magma have forced the overlying layers of rock to arch upwards.

Mons Rümker: the Moon's largest dome. Only a few of the largest domes are visible through small telescopes. The largest of these is **Mons Rümker**, a broad and distinctly lumpy plateau some 70 km in diameter that rises up to 300 m above the

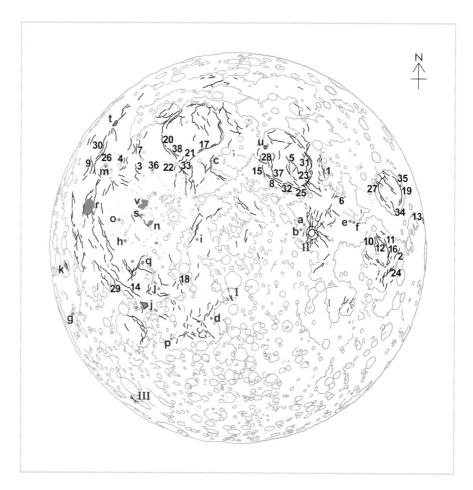

Fig. 6.5. Map showing the distribution of the Moon's wrinkle ridges and domes. Many of these are visible through binoculars (see text), but most, particularly the domes, require a telescope to observe properly.

Key: **Dorsa/dorsum – 1** Dorsa Aldrovandi **2** Dorsa Andrusov **3** Dorsum Arduino **4** Dorsa Argand **5** Dorsum Azara **6** Dorsa Barlow **7** Dorsum Bucher **8** Dorsum Buckland **9** Dorsa Burnet **10** Dorsa Cato **11** Dorsum Cayeux **12** Dorsum Cushman **13** Dorsa Dana **14** Dorsa Ewing **15** Dorsum Gast **16** Dorsa Geikie **17** Dorsum Grabau **18** Dorsum Guettard **19** Dorsa Harker **20** Dorsum Heim **21** Dorsum Higazy **22** Dorsum Lambert **23** Dorsa Lister **24** Dorsa Mawson **25** Dorsum Nicol **26** Dorsum Niggli **27** Dorsum Oppel **28** Dorsum Owen **29** Dorsa Rubey **30** Dorsum Scilla **31** Dorsa Smirnov **32** Dorsa Sorby **33** Dorsa Stille **34** Dorsum Termier **35** Dorsa Tetyaev **36** Dorsum Thera **37** Dorsum Von Cotta **38** Dorsum Zirkel.

Domes – a Arago Alpha **b** Arago Beta **c** Beer dome **d** Birt E dome **e** Cauchy Tau **f** Cauchy Omega **g** Darwin dome **h** Encke K dome **i** Gambart C dome **j** Gassendi H domes **k** Grimaldi dome **l** Herigonius dome **m** Herodotus dome **n** Hortensius domes **o** Kepler dome **p** Kies Pi **q** Lansberg D dome **r** Marius Hills **s** Milichius Pi **t** Mons Rümker **u** 'Valentine' dome **v** Domes near Mayer. **Craters with dorsa – I** Alphonsus **II** Lamont **III** Wargentin. Credit: Peter Grego surrounding plains of northern Oceanus Procellarum. Mons Rümker is best seen when it has just emerged into the lunar morning Sun; it is completely untraceable under a high Sun. Sunrise: 12d. Sunset: 26d.

Impact Craters

Many of the younger lunar craters measuring between around 30 and 100 km in diameter display all the obvious hallmarks of impact. With their sharp rims rising hundreds of meters above their surroundings, terraced inner walls, deep floors and central mountain peaks, all surrounded by extensive radial impact structure and rays, they represent most people's idea of what a "real" impact crater looks like. Notable large-impact craters of this type include Copernicus (below).

Fig. 6.6. Some prominent lunar craters and craters with bright ray systems.
Key: Prominent craters: **a** Copernicus **b** Endymion **c** Clavius **d** Ptolemaeus **e** Gassendi **f** Plato **g** Schickard **h** Petavius **i** Grimandi **j** Theophilus
Crater ray systems: **1** Anaxagoras **2** Thales **3** Aristarchus **4** Kepler **5** Copernicus rays **6** Proclus **7** Byrgius A **8** Tycho **9** Messier **10** Bessel. Credit: Peter Grego

Large lunar craters can have a depth–diameter ratio as great as 1:30, a fact that surprises anyone who has observed craters near the Moon's terminator and imagined they were incredibly deep hollows. To anyone standing on the surface of the Moon, the summit of a 3,000 m mountain would disappear below a flat horizon at a distance of around 100 km, so it would be possible to stand on the rim of many large craters and be unable to see across to the opposite side of the crater because the curvature of the Moon's surface takes it far below the horizon.

Copernicus: mighty impact crater

Of all the Moon's large, relatively fresh impact craters, **Copernicus** surely ranks as the most impressive. Copernicus and its environs are thrilling to observe, and the bright rays that spread in all directions from it can be resolved into a complex series of radial rays that extend in all directions for more than 800 km over the surrounding regions.

Copernicus arrived on the lunar scene around 900 million years ago, when an asteroid smashed into the lunar crust on the southern mountain border of Mare Imbrium. The impact vaporized the local lunar crust, carving out a 93 km crater and flinging out masses of debris, larger fragments of which produced secondary impact craters and pitting arranged in a radial fashion around Copernicus; traces of this secondary, impact structure can be seen in the extensive radial grooves and ridges surrounding the crater, although much of the finer chain craters cannot be resolved through binoculars and small telescopes. One interesting crater thought to have been formed secondary to the Copernicus impact is the keyhole-shaped conjoined craters **Fauth** (12km) and **Fauth A** (9.6km), a short distance south of Copernicus, visible through binoculars.

Copernicus's floor, 3,700 m beneath its rim, has a group of central peaks that can be discerned through binoculars – their size and configuration is somewhat similar to those of the Virgin Islands, east of Puerto Rico. Complicated terracing structure adorns Copernicus' inner walls, and much of this is resolvable, the crater's inner slopes stepping down from the rim to the floor like a massive open-cast lunar mining project. A marked "kink" – the result of a landslide – can also be observed on Copernicus's eastern rim. Sunrise: 8d. Sunset: 23d.

Some Interesting Large Lunar Craters

Endymion (125km). A prominentm dark-floored crater near the northeastern limb, easy to locate whenever it is illuminated. Nearby, on the northern border of **Lacus Somniorum** (Lake of Dreams), can be found the prominent duo of **Atlas** (87km) and **Hercules** (67km). Sunrise: 3d. Sunset: 17d.

Clavius (225km). A very large, impressive crater in the heavily cratered southern uplands. With a small telescope, a small arc of unconnected craters can be discerned on Clavius' floor. Clavius is difficult to discern when illuminated by a high Sun. Sunrise: 8d. Sunset: 23d.

Ptolemaeus (153km). A large flat floored crater just south of the centre of the Moon's disc. Ptolemaeus's southern wall is joined by **Alphonsus** (118km), a crater with a large central peak and several small, dark patches on its floor. To its southwest, the crater **Alpetragius** (40km) has a very large rounded central mountain. South of Alphonsus lies **Arzachel** (97km), a prominent deep crater with internal terracing and a central mountain massif. Under a high Sun these craters are all difficult to discern through binoculars and small telescopes. Sunrise: 7d. Sunset: 22d.

Gassendi (110km). A large crater on the northern shore of Mare Humorum whose floor contains central mountains and a number of hills, though the numerous rilles on its floor are unresolvable through small telescopes. Sunrise: 10d. Sunset: 25d.

Plato (101km). A prominent dark, flat-floored crater, sunk 2,000 m beneath the level of the lunar Alps. The shadows cast by its rim onto its floor at sunrise and sunset are fascinating to follow. Plato can easily be discerned any time it is illuminated. Sunrise: 7d. Sunset: 22d.

Schickard (227km). A vast crater in the southwestern quadrant with a flooded, distinctly patchy floor dotted with several smaller craters. Easy to locate through binoculars under high illumination. Sunrise: 12d. Sunset: 26d.

Petavius (177km). An imposing crater near the southeastern limb. It has massive, complex walls and a huge central mountain massif. Small telescopes will resolve a prominent linear rille that runs 50km across its southwestern floor from the central peak to the wall. Petavius represents something of a challenge to locate when illuminated by a high Sun. Sunrise: 2d. Sunset: 17d.

Grimaldi (222km). A large, dark circular plain near the Moon's western limb. Grimaldi occupies the central portion of a larger, less distinct basin 430 km in diameter. Grimaldi is easy to locate any time it is illuminated, even at an extreme libration. Sunrise: 12d. Sunset: 27d.

Theophilus (100km). A deep, prominent crater on the northwestern border of Mare Nectaris. Theophilus' inner walls display intricate terracin, and a large group of peaks rises above its floor. Theophilus overlies Cyrillus (98km) to its southwest, which in turn links with Catharina (100km). Together, the trio makes a superb sight under a low Sun. Theophilus' rim and central peak show up brightly under high illumination, but Cyrillus and Catharina are more difficult to identify at this time. Sunrise: 5d. Sunset: 19d.

Ray Studies

Bright ray systems cover the Moon's surface, and they are particularly noticeable around full Moon. Small instruments can be usefully employed for the study of lunar ray systems, as they may be used to survey the whole Moon or large portions of it. Craters with bright or unusual rays include the following:

Anaxagoras (51km). A prominent ray crater near the northern limb.

Thales (32km). Center of a bright, symmetrical ray system near the Moon's northeastern limb.

Aristarchus (40km). Brilliant crater in Oceanus Procellarum, the center of an asymmetric ray system traceable for 300 km to its east.

Kepler (32km). Unspectacular parent crater of a spectacular symmetrical splash of rays, fingers of which stretch to distances of 600 km.

Copernicus (93km). Giant impact crater south of Mare Imbrium. Its rays can be traced to distances of up to more than 800 km (see above for description of the crater).

Proclus (28km). Small, bright, sharp-rimmed crater west of Mare Crisium, the centre of a prominent asymmetrical ray system arranged in a broad fan shape. None of Proclus' rays cover the light gray patch of Palus Somni (Marsh of Sleep) west of Proclus.

Byrgius A (17km). A small, bright crater at the center of a prominent splash of rays, easily visible through binoculars, that reach distances of more than 300 km.

Tycho (85km). A large and very prominent crater in the southern uplands, the center of the Moon's most extensive ray system. Tycho's rays spread in a multitude of bright linear fingers of varying length, the longest of which can be traced up to 1,500 km from Tycho. Tycho's rim is surrounded by a wide, dark collar of impact melt, easily discernible through binoculars and small telescopes. Sunrise: 8d. Sunset: 22d.

Studies of ray systems using binoculars and small telescopes will enable the observer to gain an overall picture of the major lunar ray systems.

Questions: Do the ray parent craters occur mainly in highland or marial areas? Do these craters form any noticeable groups or clusters? Are the rays distributed evenly around their parent craters? Do rays from different systems overlap? Is it possible to determine which system is youngest? Are there indications that any of the rays emanate from the Moon's far side? Are the rays brightest at full Moon, or are they at maximum brightness when the Sun has culminated at their location? Are the parent craters consistently brighter than their rays, or do any of the rays exceed the brightness of their parents? In addition to the major ray systems mentioned above, what other craters possess ray systems? Are there any rays or similar features of high albedo that do not appear to be associated with any parent craters?

Recording Ray Systems

Note when the ray systems around certain craters first become visible during the lunation and when they are no longer visible under an evening illumination. Using the intensity, estimations guide (see Observing and Recording the Moon), record how the brightness of the rays changes over the lunation. Plot a curve of the apparent variation in brightness of the rays throughout the lunar day.

Use the whole Moon observing blank to record the lunar ray systems as seen at low power through your binoculars or small telescope. Mark the terminator clearly on the drawing. Use bold lines to indicate the parent craters and their bright, prominent rays, and dashed lines to indicate less prominent rays.

Features to note. Apparent source of rays or associated features; symmetry of ray system; overall pattern of rays; start and end point of rays and ray systems; tails of rays, e.g., single spikes or streaks, bright patches or stains on the lunar surface; when rays or ray systems first become visible at lunar sunrise or lost at sunset near the terminator; intensity estimates; coloration (a very subjective area depending on the observer and instrument used, visibility, etc.).

Faults

Finding Fault in the Lunar Crust

Like the crust of the Earth, the Moon's crust is capable of being deformed by stresses. Beyond a critical degree of compression or tension, the rocks suddenly fault, giving rise to fault scarps, ridges and rilles. The Moon's crust has long succumbed to all of its pent-up stresses, and no major lunar faulting has happened for billions of years. Some faults can be very deep-seated and appear to cut straight through the surface, bisecting preexisting features such as craters and mountain ranges

Fig. 6.7. Map showing the distribution of the Moon's larger rilles. Key: **Vallis** – **a** Vallis Alpes **b** Vallis Baade **c** Vallis Inghirami **d** Vallis Palitzsch **e** Vallis Rheita **f** Vallis Schröteri. **Rupes** – **I** Altai **II** Cauchy **III** Liebig **IV** Mercator **V** Recta **VI** Toscanelli. **Rima/Rimae (craters with rimae in bold)** – **1 Alphonsus 2 Arzachel 3 Atlas 4 Chacornac 5 Cleomedes 6 de Gasparis 7 Furnerius 8 Gärtner 9 Gassendi 10 Goclenius 11 Gutenberg 12 Hedin 13 Hevelius 14 Janssen 15 Palmieri 16 Petavius 17 Pitatus 18 Posidonius 19 Riccioli 20** Rima Ariadaeus **21** Rimae Aristarchus **22** Rima Birt **23** Rima Bond **24** Rima Bradley **25** Rima Burg **26** Rima Calippus **27** Rima Cardanus **28** Rima Cauchy **29** Rima Conon **30** Rima Daniell **31** Rimae Darwin **32** Rimae Doppelmayer **33** Rima Flammarion **34** Rimae Fresnel **35** Rima Gay-Lussac **36** Rimae Goclenius **37** Rima Grimaldi **38** Rima Hadley **39** Rima Hansteen **40** Rima Hase **41** RimaHesiodus **42** Rimae Hippalus **43** Rima Hyginus **44** Rimae Hypatia **45** Rimae Littrow **46** Rimae Maclear **47** Rimae Maestlin **48** Rima Marius **49** Rimae Menelaus **50** Rimae Mersenius **51** Rima Messier **52** Rima Milichius **53** Rima Oppolzer **54** Rimae Parry **55** Rimae Plato **56** Rimae Plinius **57** Rimae Prinz **58** Rimae Ramsden **59** Rimae Ritter **60** Rimae Römer **61** Rima Sharp **62** Rimae Sirsalis **63** Rimae Sosigenes **64** Rimae Sulpicius Gallus **65** Rimae Triesnecker. Credit: Peter Grego

with impunity. A few of these fault features are visible through binoculars and small telescopes:

Vallis Alpes: Giant Rift Valley of the Alps. Vallis Alpes (Alpine Valley), 130 km long and in places 18 km wide, cuts cleanly through the lunar Alps and, is visible as a dark line after local lunar sunrise and before sunset. Under a high Sun, the floor of the valley is noticeably darker than the surrounding mountains. Earth's largest land gorge, the Grand Canyon in Arizona, averages 13 km wide and 1,600 m deep. While the Grand Canyon has been hewn out of the Earth's crust by the continuous erosive action of the Colorado river over hundreds of thousands of years, erosion is not the cause of its lunar counterpart: there never have been appreciable quantities of running liquid water on the Moon's surface. In the case of the Alpine Valley, tension in the lunar Alps caused two parallel faults to appear across the mountain range, and the area bounded by the faults sank down below the level of the mountain tops. This kind of valley is known as a graben – it is by far the largest linear rille on the Moon. The floor of the Alpine Valley was later flooded with lava flowing from the maria. Sunrise: 7d. Sunset: 21d.

Rupes Recta: textbook fault in Mare Nubium. Rupes Recta (Straight Scarp) is a 110-km-long normal fault in eastern Mare Nubium caused by crustal tension. The western side of the fault has dropped by 300 m, producing a clean-cut scarp face. At sunrise, binoculars and small telescopes will reveal Rupes Recta's broad black shadow, but the bright scarp face itself, illuminated by an evening Sun, represents something of a challenge. Sunrise: 8d. Sunset: 22d.

Rupes Altai: arcuate fault of the Nectaris impact basin. Southeast of Mare Nectaris, one of the Moon's most spectacular fault features, Rupes Altai (Altai Scarp) presents a curving, scalloped cliff of staggering proportions – almost 500 km long, running from west of the crater Catharina to Piccolomini. The origin of the scarp is linked to the stresses set up in the lunar crust by the asteroidal impact that carved the Nectaris impact basin. The inner part of the basin has dropped by up to 1,000 m, exposing a scarp face along the line of a deep-seated fault. The young crescent Moon shows the Altai Mountains as a bright, winding line, in places up to 15 km wide. At sunset the scarp casts an irregular shadow onto the landscape beneath it. Sunrise: 5d. Sunset: 19d.

Libration Features

Observing near-limb features is a challenge that appeals to those interested in taking their astronomical studies into places rarely viewed by human eyes. Observing features near the limb will enable the observer to gain insight into the topography of the near-limb regions and the Moon's patterns of libration.

Mare Orientale: vast lunar bullseye. Mare Orientale (the Eastern Sea) is perhaps the most famous of the Moon's libration features. At a favorable libration, the formation may be discerned using binoculars, making a recognizable "dent" on the western edge of the Moon. Mare Orientale is located beyond the mean western lunar limb (most of the 300-km-diameter sea lying further than 90°W), and a good libration is needed to bring it into view. Mare Orientale occupies just the central part of a 930-km-diameter multiring impact basin whose outer edge is outlined by Montes Cordillera. The outer regions are stained by lava flows and, of these, Lacus Veris and Lacus Autumni (Lakes of Spring and Autumn) are situated just on the near-side and may be seen through binoculars. while Mare Orientale itself is beyond the limb.

Mare Humboldtianum. A crescent-shaped sea on the northeastern limb some 300 km across, situated wholly on the mean near side. Humboldtianum occupies the central part of a larger impact basin 650 km in diameter, which extends onto the far-side.

Mare Australe. A circular area near the southeastern limb, around 1,000 km in diameter, composed of a network of flooded, dark-floored craters and plains. Mare Australe sits more-or-less squarely on the 90°E line.

Bailly. A huge crater, 305 km in diameter, located in the libration zone near the Moon's southwestern limb. It is an old and rather eroded feature, with a low wall and rough cratered floor, but it makes a superb sight when it has emerged into the morning sunlight during a favorable libration. When devoid of shadow, Bailly is virtually invisible.

Observing Lunar Eclipses

From time to time the Moon slips into the shadow cast by the Earth into space. The shadow has two components: a dark core, the umbra, and a fainter outer region surrounding it, the penumbra. Penumbral eclipses take place when the Moon traverses the Earth's penumbral shadow. Because the penumbra has no perceptible color, appearing washed-out with an ill-defined outer edge that is impossible to detect visually, penumbral eclipses alone are pretty unspectacular. Only the most enthusiastic watcher of the Moon will bother to rise early in the morning to view this type of eclipse. Binoculars deliver the best view.

Partial lunar eclipses are more satisfying from an observational stance. The Moon first enters the penumbra, and about an hour later it partially enters the darker umbral shadow. Half an hour or so before the Moon makes first contact with the umbra, penumbral darkening can be made out at the Moon's preceding limb. At maximum eclipse, a portion of the Moon remains illuminated by direct sunlight, and the glare it causes makes the eclipsed part of the Moon appear dark and rather colorless.

A total lunar eclipse is one of astronomy's most spectacular sights. The Moon can take an hour or so to slip entirely into the umbra. At totality, although the Moon is completely covered by the Earth's shadow, it never vanishes from view altogether because a certain amount of sunlight is refracted onto the lunar surface through the Earth's atmosphere. Totality can last for nearly 2 hours during the longest lunar eclipses, when the Moon passes almost centrally through the Earth's shadow. No two total lunar eclipses appear the same; the hues, color distribution and darkness of the umbra always vary. Much depends on cloud and high-altitude dust in the Earth's atmosphere and how deeply the Moon is immersed into the Earth's umbral shadow.

Danjon Scale

A classification for total lunar eclipses, devised by André Danjon, serves as a rough guide to the observer. The Danjon scale takes into account the observed brightness of the totally eclipsed Moon and the colors of the umbra.

L = 0   Very dark eclipse. Moon almost invisible, especially at mid-totality.

L = 1   Dark eclipse, gray or brownish color. Details seen only with difficulty.

L = 2   Deep red or rust-colored eclipse. Very dark central shadow, while outer edge of umbra is relatively bright.

L = 3   Brick-red eclipse. Umbral shadow may have a bright or yellow rim.

L = 4   Very bright copper-red or orange eclipse. Umbral shadow has a very bright bluish edge.

Binoculars give the observer the best overall view of lunar eclipses. The colors can be striking, and the observer is often struck by a remarkable three-dimensional impression of the Moon floating in space against the starry background.

Line drawings on preprepared blanks may be attempted at 20-minute intervals. Accompanying these drawings, notes on the definition of the umbra's edge and any apparent irregularities in its outline can be taken, along with impressions of the darkness and coloration within the umbra. Notes on the visibility, apparent brightness and definition of certain lunar features may be taken throughout the eclipse – some bright craters like Aristarchus and Tycho may be visible throughout the entire event, distinguishable even at mideclipse. Some dark features, too, such as lunar seas or smaller, dark-flooded craters like Plato or Grimaldi may also be discerned while they are in shadow.

A good estimate of the brightness of the totally eclipsed Moon can be made using binoculars the wrong way around, comparing the brightness of the tiny condensed image of the Moon with nearby stars seen with the unaided eye.

Shadow Contact Timings

It is possible to time, to the nearest minute, the passage of the edge of the umbra over certain prominent lunar features; contact timings are made upon the feature's immersion into, and emergence from, the umbra. Through low-power binoculars, the following features can be used as shadow contact timing points (in order of immersion): Grimaldi, Aristarchus, Kepler, Copernicus, Tycho, Plato, Manilius, Posidonius, Proclus, and Langrenus.

Smaller bright features that are suitable for timings made using a small telescope include the following: Lohrmann A; Byrgius A; Mons Hansteen; Aristarchus (central peak); Encke B; Kepler; Bessarion; Brayley; Milichius; Euclides; Agatharchides A; Darney; Copernicus (central peak); Pytheas; Guericke C; Birt; Tycho (central peak); Mons Pico; Bancroft; Mösting A; Bode; Cassini A; Egede A; Manilius (central peak); Pickering; Menelaus; Dionysius; Nicolai A; Plinius; Hercules G; Censorinus; Rosse; Carmichael; Proclus; Stevinus A, and Langrenus M.

Fig. 6.8. Eclipse binocular observation. Credit: Peter Grego

Imaging Lunar Eclipses

A focal photography using binoculars and small telescopes is perfectly possible using standard compact film cameras or digicams. Because the brightness of eclipses varies so considerably, no hard guides can be given suggesting optimum exposures during totality using a standard film camera. It is best to bracket your photographs and experiment with exposures as the event unravels. The partial phases will need a little more exposure than if you were taking photographs of the Moon's regular phases, in order to reveal anything of the tones within the umbra, but undriven exposures will suffer from progressively more blurring. Digicams are a great deal more versatile than conventional cameras, and far more success can be had using a digicam hand-held to the eyepiece of a pair of binoculars or an undriven telescope than using a conventional film camera. With their sensitive CCD chips, shorter exposures mean less blurring when photographing through an undriven instrument. Images taken with digital cameras can be instantly reviewed and deleted or stored according to their quality.

Chapter 7

A Survey of the Moon's Near Side

This survey of the Moon's near-side divides the lunar disc into 16 areas. Each area is accompanied by a map showing all the main features visible through a 100 mm telescope, plus a descriptive text. The maps correspond with those published in the *Hatfield Photographic Lunar Atlas* published by Springer-Verlag (see the resources section of this book), one of the least ambiguous maps in print, drawn with clear lines and easy to use in the field at the eyepiece.

A conventional telescopic orientation, with south at the top and east to the right, has been chosen, since this corresponds with most lunar observers' telescopic view of the Moon from the northern hemisphere. Coordinates follow IAU convention, put in place in 1961, which reverses the "classical" directions of east and west. This means that Mare Orientale (the Eastern Sea), named around a century ago, now officially lies in the western hemisphere of the Moon. Confusion can arise because some authors continued to employ classical coordinates long after the IAU ruling, often making their descriptions of lunar features difficult to understand.

The maps are all drawn to the same scale and take in an area of around 8×8 arc-minutes (with the exception of the four far quadrant maps – 4, 8, 12 and 16 – which are to the same scale but cover less area). This roughly corresponds to the area visible through a telescope eyepiece with a 50° apparent field at a magnification of around ×380. The northeastern quadrant is covered by maps 1–4, and maps 5–8 cover the northwestern quadrant; the southwestern quadrant is covered by maps 9–12, and maps 13–16 cover the southeastern quadrant.

This descriptive survey of the Moon details most of the features shown on the maps, in addition to a number of smaller or subtler features that may not be indicated. Finer lunar features and the telescope size required to resolve them are described in detail. In many instances, features (or groups of features) of interest that lie on, or extend beyond, the edges of the maps may be mentioned, but there is minimal repetition in the descriptions of adjacent areas. The division between each area described does not follow the exact lines of a grid, as this would contrive to split up the descriptions of some features and associated groups of features between areas. The Moon's topography wasn't planned along the neat lines of a modern city! The descriptive text in each section covers only features that are visible on each map. As a general rule, the survey describes features (or groups of features) in each map section from the north to the south, and (where applicable) features that lie within the libration zones are described last. Each feature's official IAU name (and in some cases its unofficial name) is given along with the English

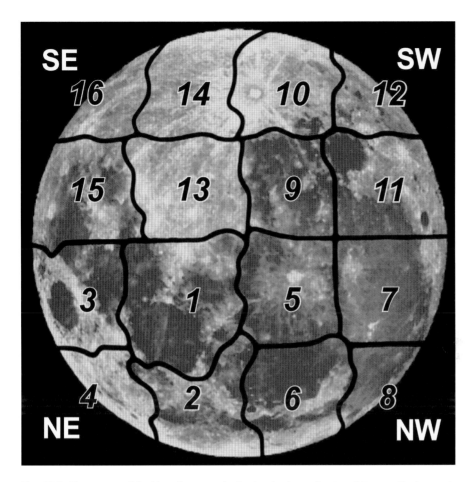

Fig. 7.1. This survey of the Moon's near side divides the lunar disc into 16 areas Each area is accompanied by a map showing all the main features visible through a 100mm telescope, plus a descriptive text The maps correspond with those published in the *Hatfield Photographic Lunar Atlas* Springer (see Resources), among the least ambiguous maps in print, drawn with clear lines and easy to use in the field at the eyepiece. A conventional telescopic orientation with south at top, east to the right, has been chosen, since this corresponds with most lunar observers' telescopic view of the Moon from the northern hemisphere Co-ordinates follow IAU (International Astronomical Union) convention. The maps are all drawn to the same scale, and take in an area of around 8 × 8 arcminutes (with the exception of the four far quadrant maps, 4, 8, 12 and 16, which are at the same scale but cover less area) The northeastern quadrant is covered by maps 1–4, and maps 5–8 cover the northwestern quadrant; the southwestern quadrant is covered by maps 9–12, and maps 13–16 cover the southeastern quadrant. Credit: Peter Grego

language meaning of the name (where applicable) and the feature's dimensions. Metric measurements have been used throughout. Bold letters indicate the first mention of a feature in the main descriptive text.

Some Lunar Terminology

A number of Latin names on the Moon have survived the centuries. The IAU has extended some of these terms to describe similar features observed on other solid worlds within the Solar System.

Albedo feature	An area distinguished by its reflectivity
Catena	A chain of craters
Cleft	A small rille
Crater	A circular feature (sometimes depressed below the surface level)
Dorsum (dorsa)	Ridge
Ghost crater	A crater that is barely visible because of flooding or erosion
Lacus	"Lake" – a small plain
Landing site name	Lunar features at or near Apollo landing sites
Mare (maria)	"Sea" – a large plain
Mons (montes)	Mountain
Oceanus	"Ocean" (Oceanus Procellarum) – a very large marial plain
Palus	"Swamp" – a small plain
Planitia	Low plain
Promontorium	"Cape" – a headland
Rima (rimae)	Rill
Rupes	Scarp
Sinus	"Bay" – a small plain
Statio	"Base" (Statio Tranquillitatis)
Vallis	Valley

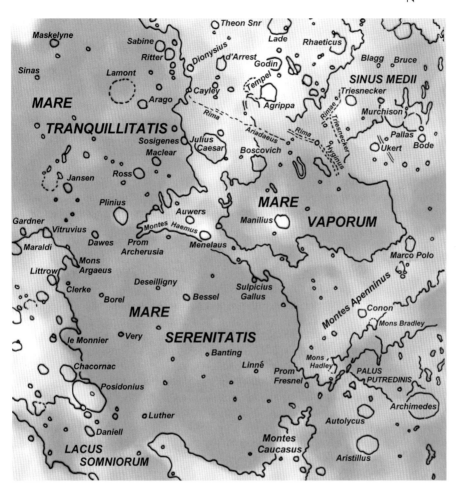

Northeastern Quadrant

Area One

This area is largely occupied by marial plains. Our survey includes the whole of **Mare Serenitatis** (Sea of Serenity) and **Mare Vaporum** (Sea of Vapors) and encompasses the western sector of **Mare Tranquillitatis** (Sea of Tranquil ity). Notable topographic features include the eastern part of the **Montes Apenninus** (Apennine Mountains), the prominent crater Posidonius, the unusual wrinkle feature Lamont and several major rille systems. The sunrise terminator crosses this area between around 4–8 days, and the sunset terminator between around 19–23 days.

Mare Serenitatis is a well-defined lunar sea with a roughly circular outline. It has an average width of 630 km, but, measured from its northwestern shore to its southeastern margin, it has a maximum breadth of 700 km. The teardrop shape of Mare Serenitatis is caused by the striking angularity of its northwestern border. Having a total surface area of a little more than 300,000 sq km, Mare Serenitatis is the sixth largest marial expanse on the entire Moon, and it presents an easy object to see without any optical aid.

Mare Serenitatis has one of the most tonally varied surfaces of all the lunar maria, and this is most noticeable when it is illuminated by a high Sun at around 12 days. This tonal variation is produced by the different albedo of the numerous lava flows making up the mare, in addition to patches of overlying bright material ejected from impact craters, some of which lie beyond the borders of the mare. Binoculars will easily show mottling within Mare Serenitatis, and a telescope will reveal that the patchiness tends to be very sharply delineated in places. The northwestern and western borders of the sea are stained with dark patches broken up by numerous light rays, some of which appear to emanate from the impact craters **Aristillus** and **Autolycus**, a couple of hundred kilometers to the west (see Area Three), across the mountain border in the adjoining **Mare Imbrium** (Sea of Rains – see Area Six). Under a high Sun, the neat angular arrangement of these rays rather resembles a large airport's landing strips. Mare Serenitatis's southern and eastern borders have clear-cut dark lava patches that link with the lavae of **Mare Tranquillitatis** to the immediate south. There is a particularly intense patch of dark lava nearby in a small valley between two prominent mountains, **Mons Argaeus** (50 km long, 2,500 m high) and **Mons Vitruvius** (30 km long, 1,550 m high); this valley was the landing site of Apollo 17 in December 1972. Several long, light-colored ray fingers stretch northwards across Mare Serenitatis, the most prominent of which stretches for around 300 km, emanating from the bright crater **Menelaus** (27 km) on the southern border of the mare, through crater **Bessel** (16 km) and across to the small crater **Bessel D**.

Under a low angle of illumination, Mare Serenitatis displays an intricate system of wrinkle ridges that snake their way around the sea, keeping roughly parallel to its border. The most prominent of these wrinkle ridges is **Dorsa Smirnov**, an impressive braided "rope" of wrinkles, in places 15 km wide, about 100 km from the mare border. Dorsa Smirnov is around 200 km long, and can easily be glimpsed through a small telescope when it is near the terminator. It originates at a point around 50 km west of Posidonius and flows south through the small crater **Very** (5 km), south of which the ridge bends and narrows. Another ridge rises again a little further south, where it is called the **Dorsa Lister**, a system of ridges around 300 km long. Dorsa Lister curves parallel to the southern mare border, where it is encroached upon at right angles by a number of narrower ridges. One narrow, plaited ridge stretches to Dorsa Lister from the crater **Dawes** (18 km), some 110 km away, and **Dorsum Nicol** (50 km long) can be seen further west. Dorsa Lister continues to track north, where it meets the flanks of the crater Bessel. Some low ridges proceed north of Bessel, and these meet up with **Dorsum Azara** (110 km long) near the center of Mare Serenitatis. Near the western shore of Mare Serenitatis are a number of narrower dorsa, including **Dorsum Buckland** (150 km long), **Dorsum Gast** (60 km long), **Dorsum Owen** (50 km long) and **Dorsum von Cotta** (220 km long). North of Dorsum Owen, 100 km from the western shore, is a bright circular patch 10 km wide that surrounds a tiny 2.4-km-diameter crater called **Linné**. One of the Moon's largest single domes lies 90 km northwest of Linné.

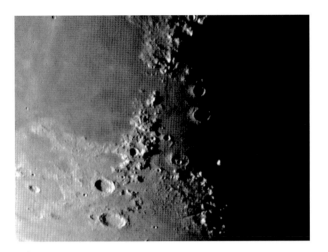

Fig. 7.2. Montes Caucasus.
Credit: Brian Jeffrey

It is an unnamed, oval-shaped elevation (35 × 25 km) whose surface is punctured by several small sharp crags; its unofficial name is the "Valentine Dome".

Mare Serenitatis has a discontinuous mountain border. The far-northwestern border is marked by **Montes Caucasus** (Caucasus Mountains), an impressive range that makes a north–south wedge more than 500 km long, separating eastern Mare Imbrium and northwestern Mare Serenitatis. Peaks in the range reach heights of 6,000 m in places. The area around 270 km of the northern border of Mare Serenitatis is marked by a broad range of unnamed hills. A gap of 150 km separates these hills from the prominent crater Posidonius. At 95 km across, Posidonius lies at the northeastern border of Mare Serenitatis at the entrance to **Lacus Somniorum** (Lake of Dreams; see Area Two). A 100mm telescope will reveal the complex nature of Posidonius's floor. A prominent clear cut bowl-shaped crater **Posidonius A** (12 km) sits slightly west of center. To its east there are a number of linear rilles, the **Rimae Posidonius**, the most prominent of which cuts 50 km across the center of the floor, and it is cut across itself at a right angles by a smaller rille. Another, longer, rille makes its way along the floor near the inner western wall. The best views of the Rimae Posidonius are to be obtained through a 150mm telescope at high magnification. The eastern part of Posidonius' floor is crossed by a prominent curving ridge – a huge block of crust that has slipped away from the main wall. Posidonius's northern rim is dented by a chain of smaller craters, **Posidonius J, B** and **D**. Under a high Sun, Posidonius remains clearly visible, with a bright rim and a light-colored floor. Adjoining the crater's southern wall is the disintegrated old crater **Chacornac** (51 km). Close telescopic scrutiny will reveal that, like Posidonius, Chacornac has an off-center bowl-shaped crater and a system of small rilles that bisect its floor.

A western extension of **Montes Taurus** (see Area Three) marks the eastern border of Mare Serenitatis. Ninety kilometers south of Posidonius, a prominent bay is formed by the flooded crater **le Monnier** (61 km), site of the robot lunar crawler Lunokhod 2's excursion in 1973. The flat-floored crater **Littrow** (31 km) is sunk into the mountains, and to its north and west can be found **Rimae Littrow**, a complex of narrow rilles that extend from the mountains into the mare itself around the little crater **Clerke** (7 km). These rilles require a 150mm telescope to resolve. The large headland of Mons Argaeus looks westwards across the marial

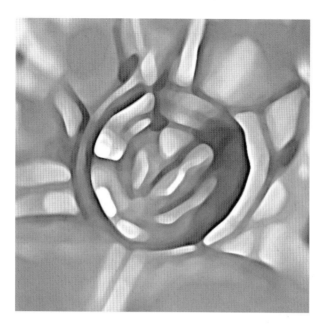

Fig. 7.3. Observational drawing of the crater Plinius. Credit: Peter Grego

plain between Mare Serenitatis and Mare Tranquillitatis towards Dawes and **Plinius** (43 km), the latter being a prominent polygonal crater with a nicely wrinkled collar and a blocky floor. To its north can be found **Rimae Plinius**, a set of three linear rilles, each around 100 km long, that look like cats' claw marks. The rilles lead across to where Mare Serenitatis' mountain border resumes, in the form of **Montes Haemus**, beginning with the crenulated headland of **Promontorium Archerusa**. Montes Haemus are one of the Moon's less grand-looking named mountain ranges, a narrow, knobbly plateau some 400 km in length that marks the southwestern boundary of Mare Serenitatis. Made up of rounded elevations, its highest summits reach around 2,000 m. The bright crater, Menelaus, sits in the mountains at the southern edge of the sea. To the west, the mountains take on a markedly striated appearance, with ridges and valleys radial to Mare Imbrium; their sculpting (along with other features radial to Mare Imbrium elsewhere on the Moon) is a direct consequence of the powerful explosive forces released when the Imbrium basin was created by a major asteroidal impact. Among the Montes Haemus can be found a number of sizable dark-lava plains. Proceeding westward from Menelaus, they are **Lacus Hiemalis** (Winter Lake), **Lacus Lenitatis** (Lake of Tenderness), **Lacus Gaudii** (Lake of Joy), **Lacus Doloris** (Lake of Suffering), **Lacus Odii** (Lake of Hate) and **Lacus Felicitatis** (Lake of Happiness). **Boscovich** (46 km) and **Julius Caesar** (90 km) are among several more dark-floored lakes (some of them unnamed) to be found in the mountains further south. Binoculars will reveal the patchiness of this region when it is lit by a high Sun.

The western heights of Montes Haemus mingle with the eastern part of **Montes Apenninus**, the Moon's largest and most imposing mountain range, 600 km in length and forming a spectacular sweeping border to southeastern Mare Imbrium. Montes Apenninus reach heights in excess of 5,000 m in giant angular mountain blocks such as **Mons Ampère** (30 km long), **Mons Huygens** (40 km long), **Mons Bradley** (30 km long), **Mons Hadley Delta** (20 km long) and **Mons Hadley** (25 km long) that project northwards, creating a jagged sawblade that cuts into the shore

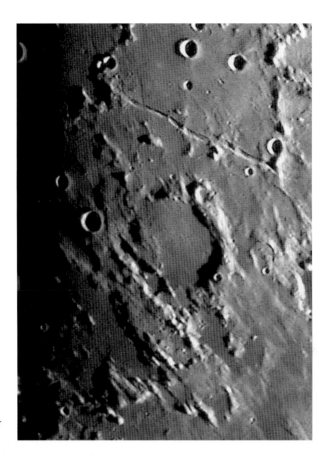

Fig. 7.4. The Julius Caesar area. Credit: Mike Brown

of Mare Imbrium – a truly superb sight through any-sized instrument when they are observed at a morning illumination, as they cast long shadows westwards across Mare Imbrium. Under a high illumination, the mountains can be seen as a prominent chain of bright points. The sharp crater **Conon** (22 km), nestled in the mountains near Mons Bradley, is also a prominent bright spot under a high Sun. The dark area between the northern Apennine mountain border of Mare Imbrium and the large crater **Archimedes** (see Area Two) is called **Palus Putredinis** (Marsh of Decay); at 180 km wide, it can just about be discerned without optical aid by those with exceptional eyesight.

    Rima Hadley, a narrow, sinuous rille 80 km long, can be found cutting across the marial plain at the base of Mons Hadley Delta. This was the site of Apollo 15's landing in July 1971. Rima Hadley is a difficult feature to observe because of its proximity to the mountains and the shadows cast by them, and the best opportunity to glimpse it takes place when it is illuminated by a Sun low in the west at local sunset. It requires a 150mm telescope to resolve. Somewhat easier to discern is the nearby **Rima Bradley** (130 km long), which cuts a curving furrow across the hills north of Mons Bradley. Several more rilles lie near **Promontorium Fresnel** at the northern tip of the Montes Apenninus. There is a 50-km-wide break in the western mountain border of Mare Serenitatis, where it links with Mare Imbrium between the northern tip of Montes Apenninus and the southern peaks of Montes Caucasus.

Mare Vaporum (Sea of Vapors) southwest of Mare Serenitatis, is a relatively smooth, dark plain around 230 km in diameter that occupies an area of around 55,000 sq km. It is visible to keen-sighted people without any optical aid. The prominent crater **Manilius** (39 km), with its sharp polygonal outline, terraced inner wall and central peaks, lies on Mare Vaporum's northeastern shore. Manilius' rim has a high albedo and it is prominent at local midday. A faint ray from Manilius projects to the west, bisecting the sea. In contrast to Manilius, an exceptionally eroded, elongated crater called **Marco Polo** (28 × 21 km) lies on the other side of Mare Vaporum. It is virtually indistinguishable from many other similar denuded features in the vicinity. The southern part of Mare Vaporum in the vicinity is hilly and striated, stained with darker patches that are aligned radial to Mare Imbrium. An unusual, unnamed feature just north of the crater **Hyginus** appears as an ill-defined crater, 20 km in diameter, with a domed floor and a central crater-let. This feature is known unofficially as "Schneckenberg" (Snail Mountain), owing to its alleged spiral structure, rendered visible only at low angles of illumination. Hyginus (10.6 km) is a keyhole-shaped crater that lies at the center of **Rima Hyginus**, one of the Moon's loveliest rilles, visible through a 60mm refractor at medium to high power. Rima Hyginus extends for more than 110 km on either side of Hyginus; one branch cuts to the northwest, the other makes its way roughly eastwards. High-magnification scrutiny through a 200mm or larger telescope will reveal much of the rille to be made up of an interlocked chain of craters. In the east

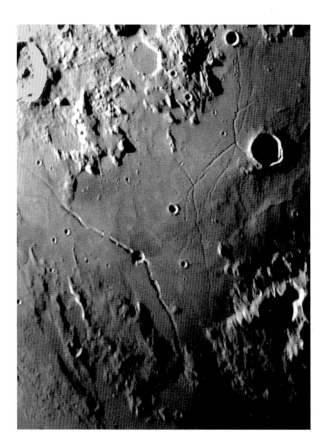

Fig. 7.5. Rima Hyginus.
Credit: Mike Brown

Fig. 7.6. Observational drawing of the crater Agrippa. Credit: Peter Grego

it transforms into a regular linear rille, which forks and fades into the level of the plain.

Hyginus is located in a northeast extension of **Sinus Medii** (Central Bay), a small, irregular-shaped mare near the center of the lunar disc that links up with Mare Vaporum to the north. Sinus Medii averages 350 km in diameter and has an area of 52,000 sq km. A magnificent system of rilles centered around the crater **Triesnecker** (26 km) sprawls north–south across Sinus Medii a short distance south of Rima Hyginus. **Rimae Triesnecker** is made up of a dozen or more sizable linear rilles that intertwine to create the finest network of valleys on the Moon; if the largest components were placed end to end, they would have a total length of around 1,000 km. To the north of Sinus Medii, **Pallas** (50 km) and **Murchison** (58 km) form a linked, but somewhat eroded, crater duo. On the opposite side of Sinus Medii, **Rhaeticus** (45 km) is a similarly eroded crater. To its east 150 km are the more sharply defined craters **Agrippa** (46 km) and **Godin** (35 km), similar-looking, somewhat triangular in outline with knobbly floors and small central peaks. An interesting looking unnamed mountain to the west of Godin, near the eroded crater **Dembowski** (26 km), appears to be a rounded plateau bisected by a straight ridge.

Rima Ariadaeus runs east across the undulating terrain between Agrippa and Julius Caesar to meet up with the border of Mare Tranquillitatis to the east. At 220 km long, and in places 9 km wide, it is one of the Moon's most impressive linear rilles. Its path is interrupted in a number of places, notably by a large mountain spur that protrudes from the bowl-shaped crater **Silberschlag** (13 km). Rima Ariadaeus is just visible through a 60mm refractor, and it would be more prominent were it angled more favorably to the Sun. The little double crater **Ariadaeus** (11 km), after which it is named, is located at its far eastern end, on the shore of Mare Tranquillitatis.

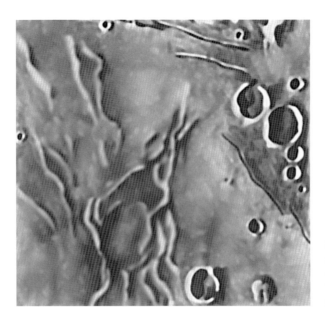

Fig. 7.7. Observational drawing of the wrinkle feature Lamont in Mare Tranquillitatis. Credit: Peter Grego

The western side of Mare Tranquillitatis is exceptionally interesting, although the features that interest the lunar observer may not be at all obvious to the casual viewer. Under a high Sun, the mare is broken into a number of well-defined units of different albedo, and this can be discerned through binoculars. Under a low angle of illumination, a vast network of low wrinkle ridges measuring more than 300 km from north to south converges on a central, roughly circular system of ridges called **Lamont** (75 km). Unique on the Moon, Lamont is best observed when it is near the 5-day-old terminator and illuminated by a rising Sun. **Arago** (26 km), a crater with a prominent mountain spine on its northern floor, lies just northwest of Lamont, and two large, knobbly domes – **Arago Alpha** and **Arago Beta** – lie north and west of Arago, respectively. Through a 150mm telescope, an unnamed line of three small domes can be observed further north of Arago Alpha. A system of fine parallel rilles, resolvable through a 150mm telescope, can be observed close to the western border of Mare Tranquillitatis. They include the widely separated duo of **Rimae Maclear** (100 km long), the parallel trio of **Rimae Sosigenes** (150 km long) and the several components of **Rimae Ritter** (100 km long) near the crater duo of **Ritter** (31 km) and **Sabine** (30 km) in the southwest corner of the mare. Apollo 11, the first manned lunar landing, touched down about 100 km east of Sabine on the southern plains of Mare Tranquillitatis in July 1969. The site has been named **Statio Tranquillitatis** (Tranquility Base) and three very small craters, **Aldrin** (3.4 km), **Collins** (2.4 km) and **Armstrong** (4.6 km), lie just to its north; they require at least a 150mm telescope to resolve.

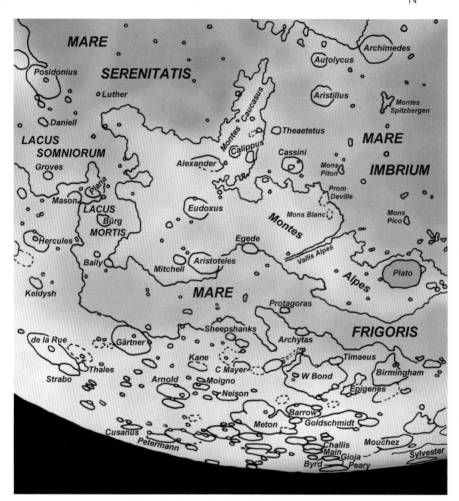

Area Two

This area, covering a large portion of the north of the Moon, including the north polar region, takes in the eastern half of **Mare Frigoris** (Sea of Cold) around to **Lacus Somniorum** (Lake of Dreams) in the east, and includes the far eastern part of Mare Imbrium and the eastern heights of **Montes Alpes** (Alps). Notable features include a number of large, foreshortened craters near the northern limb, the mighty **Vallis Alpes** (Alpine Valley), and the prominent trio of craters **Archimedes**, **Aristillus** and **Autolycus** in Mare Imbrium. The large craters **Aristoteles** and **Eudoxus** lie near the center of the region. The sunrise terminator crosses this area between around 4–days, and the sunset terminator between around 19–23 days.

As with all near-limb regions, the observability of features around the Moon's north polar region is subject to the effects of libration. The north polar region is a crater-strewn highland, and, when first sighted through the telescope eyepiece, the task of identifying features here may seem a daunting, if not near-impossible task. However, using a good telescope at a reasonably high magnification, a map and plenty of patience, the observer will soon find that the craters here don't all look the same: there are plenty of individual craters of note that can be used as navigational landmarks. On the northern shoreline of Mare Frigoris, due north of **Plato** (see Area Six) is the irregularly shaped, disintegrated crater **Birmingham** (92 km), a feature with low, roughly hewn walls, identifiable under a low illumination. Immediately northeast of Birmingham is the clear-cut crater **Epigenes** (55 km), its eastern wall abutted by a high part of the wall of the large, flat-floored crater plain **W Bond** (158 km), whose irregular southern flanks include polygonal crater **Timaeus** (33 km) and form part of the shoreline of Mare Frigoris. Several light-colored linear rays cross W Bond; these can be traced northwest to their point of origin, the brilliant-rayed crater **Anaxagoras** (51 km). Anaxagoras is the "Tycho of the north" – the most prominent landmark in the north polar region, a deep crater with a sharp circular rim. Its walls cast prominent shadows at a low illumination. Under a high Sun, Anaxagoras and its rays are easy to see – indeed, the brightening they cause can even be discerned with the unaided eye. Through binoculars, the rays can be seen to cover much of the north polar highlands and they stretch for distances of up to 600 km. Anaxagoras sits squarely over the western wall of the large, smooth-floored crater **Goldschmidt** (120 km). The actual lunar north pole lies about 480 km due north of Goldschmidt's northern wall. Several other large, smooth-floored craters are to be found to the east. **Barrow** (93 km) adjoins Goldschmidt's eastern wall. A small telescope will show its unusual wavelike southern outline. Barrow itself intrudes upon the southwestern part of **Meton** (122 km), a flat-floored crater with a well-defined rim that is indented with a number of large bays. Further east lies the circular crater **Neison** (53 km), with its flat, dark-toned floor, and to Meton's northeast is **Baillaud** (90 km), a single 14 km crater standing proudly near the center of its floor. Well-defined **Euctemon** (62 km) joins Meton's northern flanks. **Scoresby** (56 km), a deep, clean-cut circular crater northwest of Meton, is one of the most easily identifiable craters in the vicinity of the north pole. Scoresby can be seen near the limb, often full of shadow at times when surrounding features have faded in the sunlight. Scoresby is the southernmost of a line of sizable, well-preserved craters that leads north to the pole: the joined duo of **Challis** (56 km) and **Main** (46 km), **Gioja** (42 km), **Byrd** (94 km) and **Peary** (74 km). The north pole itself lies just beyond Peary's northern rim. Of course, all of these craters are greatly affected by libration, and appear exceedingly foreshortened even at a good libration. Gioja, Byrd and Peary can themselves be taken beyond the limb at an unfavorable libration. Beyond the north pole lies the large crater **Rozhdestvenskiy** (178 km), which can occasionally be glimpsed at the limb. More frequently observable is the crater **Hermite** (110 km), due west of the pole.

The dusky, gray tract of Mare Frigoris can be seen without optical aid by those with average eyesight, but it can be more difficult to discern with certainty when there is a strong libration favoring the southern limb. Mare Frigoris stretches a good way across the northern latitudes of the lunar disc, from around longitude 45°W to 45°E, a distance of almost 1,500 km. In the east, Mare Frigoris is connected to the broad, dark plain of **Lacus Mortis** (Lake of Death). The eastern half of Mare Frigoris is generally smoother than the west, with only a few minor wrinkle ridges (see Area Six for a description of the western half). A highly eroded, irregular crater, **de la Rue** (136 km), at the far eastern end of the mare, contrasts with the sharp-rimmed craters **Strabo** (55 km) and **Thales** (32 km) that adjoin it. Thales is the center of one of the Moon's brighter ray systems, and its rays mingle with those surrounding Anaxagoras, some 720 km to the west.

While the northern border of Mare Frigoris remains at a latitude of around 60°N, its southern margins plunge southwards to form a deep, broad south-pointing triangle parallel to the northeastern border of Mare Imbrium on its western side. The plains of eastern Mare Frigoris are punctuated by a number of small, but prominent, craters, notably **Archytas** (32 km), **Protagoras** (22 km), **Sheepshanks** (25 km) and **Galle** (21 km). **Democritus** (39 km), a deep, prominent crater with a central peak, lies immediately northwest of **Gärtner** (102 km), an incomplete crater that forms a semicircular bay on the northeastern shore. **Gärtner's** floor is notable for **Rima Gärtner**, an arc-shaped rille visible through a 100mm telescope. A far finer rille – one that is a challenge to resolve through a 200mm telescope on nights of good viewing – is **Rima Sheepshanks** (200 km long), one of the Moon's lengthiest single linear rilles, which cuts across the mare west of **Gärtner** to a point south of Sheepshanks.

Fig. 7.8. Craters Aristoteles and Eudoxus. Credit: Mike Brown

Aristoteles (87 km) and Eudoxus (67 km) form a prominent duo striking to observe through any telescope. The pair are visible whenever they are illuminated by the Sun, even through binoculars, since both take the form of well-defined light-colored rings under a high illumination. Aristoteles is the more impressive of the two. It has a slightly polygonal outline and broad inner walls that display some of the most extensive terracing in any crater on the Moon. The crater's floor is depressed below the mean level of the surrounding terrain, and through a 100 mm telescope it appears relatively smooth, apart from two mountain peaks that protrude from its southern floor. Aristoteles' rim is clear-cut, and displays a scalloped effect (seen in many other large-impact craters of a similar size) caused by large units of rock that have broken away from the wall and slid down it to some extent. Under morning or evening illumination, a broad flange of impact structuring can be discerned in the crater's vicinity, this taking the form of a mass of radial ridges that extend from the rim to distances of up to 100 km. Buried within this structure on the crater's eastern wall is Mitchell (30 km), a crater that predates Aristoteles and a good example of a smaller crater that is overlapped by a larger crater. South of Aristoteles the terrain gets somewhat rougher. Some 100 km to the south, Eudoxus makes an interesting near-neighbor. Although it resembles Aristoteles, close examination will show a number of subtle differences; while its internal terracing is slightly less complex, its floor is somewhat blockier. The impact structuring around Eudoxus is less grand, partly because of the preexisting

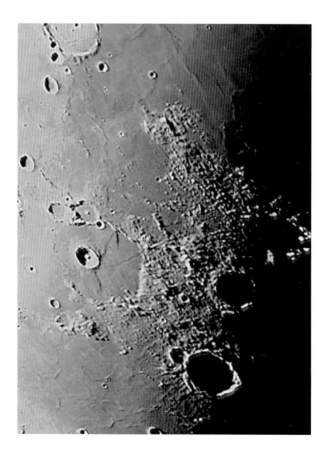

Fig. 7.9. Crater Burg in Lacus Mortis. Credit: Mike Brown

rougher nature of the surrounding terrain, with more concentric rather than radial structure being evident.

Lacus Mortis, one of the strangest-looking parts of the Moon, is located 125 km due east of Eudoxus. Lacus Mortis is the remains of a large, flooded crater 150 km in diameter. Its western wall makes a clear bay in the uplands, and this extends into narrow fingers of hills that mark the ancient crater's original rim in the north and the south. The ring is broken by lava flows to the east, but a line of hills can be seen protruding above the plain to the east. An imposing crater, **Bürg** (40 km), lies just off-center in Lacus Mortis, and this sits upon a triangular wedge of higher ground that may be Lacus Mortis' original central uplift. The western half of Lacus Mortis is crossed by **Rimae Bürg**. The most prominent rille cuts across the plain and links the southwestern wall to the central hills, and another runs from the southern wall to the midpoint of the main rille. These can be resolved through a 100 mm telescope, but other, narrower rilles in the system require at least a 150 mm telescope to discern. Conjoined craters **Plana** (44 km) and **Mason** (37 km) are depressed into the disintegrated terrain on the southern border of Lacus Mortis.

Lacus Somniorum is a sizable marial patch with a highly irregular border. About 500 km wide, it covers an area of about 70,000 sq km south of Lacus Mortis, between northeastern Mare Serenitatis and the highlands further east. Numerous low, rounded hills and domes dot the surface of Lacus Somniorum, but one of its most interesting features is **Rimae Daniell**, a system of linear rilles, the longest of which cuts 160 km east–west across the plain. A 150 mm telescope will resolve it. To the east of Lacus Somniorum can be found the disintegrated, flooded crater **Hall** (39 km), and to its south the bright, sharp polygonal crater **G Bond** (20 km). **Rima G Bond** cuts a 150 km furrow across the plain and through the hills to the west, visible through a 100mm telescope.

The magnificent mountain range of **Montes Alpes** (see map 6) commences to the west of Egede (37 km), a flat-floored ghost ring 80 km west of Aristoteles. Montes Alpes make a spectacular sight at around first and last quarter phase, whatever the size of instrument they are observed with. Montes Alpes form a vast plateau, about 350 km in length, with an area of about 70,000 sq km, and they mark the far northeastern border of Mare Imbrium. Montes Alpes and the mountains along the northern borders of Mare Imbrium do not display an obvious structuring that is radial to the Imbrium impact basin, quite unlike the topography of the highlands that form the eastern and southern borders of Mare Imbrium. Montes Alpes are divided in two by **Vallis Alpes**, one of the Moon's largest fault valleys (see below). East of Vallis Alpes, Montes Alpes is largely hilly, speckled with innumerable small mountain peaks. To the west of Vallis Alpes, and along the border of Mare Imbrium, the mountains rise to heights of more than 3,600 m. Notable individual mountains in the range include **Mons Blanc**, **Promontorium Deville** and **Promontorium Agassiz**, all rising dramatically above the shoreline of Mare Imbrium. Promontorium Agassiz marks the southernmost tip of the range. A number of equally impressive peaks can be found on the other side of Vallis Alpes, but none of these are named.

Vallis Alpes is a remarkable example of a graben, a lunar rift valley – a scaled-up version of a linear rille – that cuts through 130 km of the Montes Alpes from Mare Frigoris to Mare Imbrium. In places Vallis Alpes is 18 km wide, and its steep walls rise to an average height of 2,000 m above its floor. The valley's floor is smoother and darker than the surrounding mountains because it has been flooded with lava. A 150mm telescope will resolve parts of a small, sinuous rille in the center of the

Fig. 7.10. General view of
the Montes Alpes, showing
Plato and Vallis Alpes. Credit:
Chris Dignan (8)

valley floor, a feature that was carved by the erosive action of fast-flowing lava from
Mare Imbrium. From the plains of Mare Imbrium, the entrance to the valley begins
at a V-shaped inlet that narrows to a mountain pass just a few hundred meters
wide. After 10 km the valley opens out into a diamond-shaped "plaza", and it is at
the entrance to this plaza that the narrow medial rille begins its journey. The plaza
narrows to about 5 km, but afterwards the valley gradually widens to its maximum
width about halfway along its length. Finally, it narrows to about 7 km wide at its
junction with Mare Frigoris. As it emerges from the morning terminator, Vallis
Alpes appears as a black line, but much of the floor can be discerned at about one
day after local sunrise. In the mountains west of Vallis Alpes are the numerous
rilles of **Rimae Plato**, thought to have been cut by lava flows just like Vallis Alpes'
medial rille. The largest of these require at least a 100 mm telescope to resolve.

Montes Caucasus (see Area One), a prominent spine of mountains, stretches for
500 km southwards from the vicinity of Eudoxus to mark part of the far-eastern
border of Mare Imbrium. Tucked away in the northeastern corner of Mare
Imbrium, on the plain between Montes Caucasus and Montes Alpes, lies Cassini
(57 km), an ancient but well-defined crater whose interior has been flooded and
later dented with two sizable impact craters. The largest, **Cassini A** (17 km), lies just
north of center on Cassini's floor and has a large interior mound. The outer flanks
of Cassini somewhat resemble those "mud splash" craters on Mars, but the reason
for their rounded appearance in Cassini's case lies in the fact that it is a very

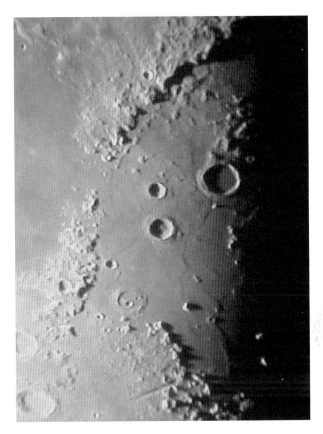

Fig. 7.11. CCD image of eastern Mare Imbrium at sunrise, with the crater trio Archimedes, Aristillus and Autolycus. Credit: Peter Grego

ancient crater, predating the Imbrium basin impact, and there has been considerable lava flooding in the vicinity. Perhaps surprisingly for such an old eroded crater, Cassini's narrow rim can easily be discerned under a high Sun. Towering at a height of 2,250 m above Mare Imbrium, 100 km due west of Cassini, the isolated

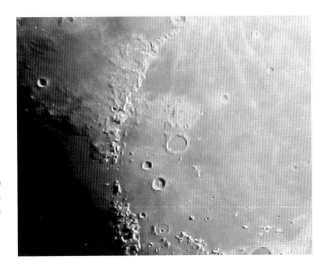

Fig. 7.12. CCD image of eastern Mare Imbrium approaching sunset, with the crater trio Archimedes, Aristillus and Autolycus. Credit: Peter Grego

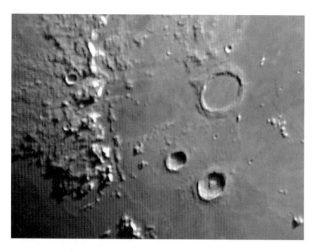

Fig. 7.13. CCD image of eastern Mare Imbrium approaching sunset, with the crater trio Archimedes, Aristillus and Autolycus. Credit: Peter Grego

mountain **Mons Piton** casts a sharp elongated shadow around sunrise and sunset. It appears as a brilliant L-shaped squiggle under a high angle of illumination.

Mare Imbrium's eastern reaches are of a lighter tone than the rest of the sea. Through binoculars it will be seen that much of the brightness is made up of ray material ejected from the impact craters **Aristillus** (55 km) and **Autolycus** (39 km), along with rays from **Copernicus** in the southwest (see Area Five) and the uplands of **Montes Archimedes** to the southeast (see Area Five). A prominent dark patch in southeastern Mare Imbrium is the lava plain Palus Putredinis (see Area One). Aristillus is a splendid crater, slightly polygonal in outline, with a broad inner wall that displays well-developed terracing, and a nice compact cluster of central mountains that rise above a smooth floor. The northeastern inner wall is marked by a prominent dark band that extends from the crater's floor to its rim. This feature, one of the most noteworthy dark albedo bands to be found in any lunar crater, does not appear to be associated with any topographic formation. Another, fainter radial band marks the northwestern interior wall. Because Aristillus lies on a generally smooth and level lava plain, it possesses one of the clearest observable systems of impact structures on the Moon for any crater of its size. In addition to radial ridges and furrows that stretch up to 50 km from the crater's rim, visible only under a low illumination, a delicate but somewhat faded ray system can be traced in all directions from Aristillus across Mare Imbrium, and even over Montes Caucasus into northern Mare Serenitatis (see Area One) under a high Sun. A curious bright apron extending 20 km is formed by the rays immediately north of the crater. Autolycus, 45 km south of Aristillus, is a smaller impact crater that has a rough, hilly floor and much less prominent external impact structuring than its neighbor.

Archimedes (83 km) is a superb crater. With is flat, lava-flooded floor and its extensive, well-structured walls, it dominates the plains of southeastern Mare Imbrium. Its floor is striated with several light rays that emanate from Autolycus to its east, and, apart from three very tiny craters near its inner wall, Archimedes's floor appears flat and topographically featureless even at very low angles of illumination, using a 150mm telescope with a high magnification. Internal lava flows have obliterated any central elevations that it may once have had. The shadows cast by Archimedes' rim onto its floor when the crater is near the terminator are fascinating to observe. In the early morning, at least seven individual spikes of shadow

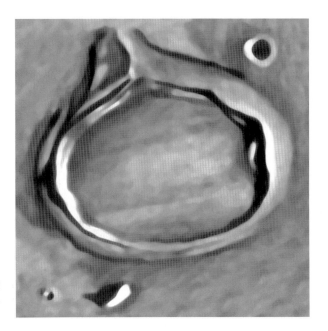

Fig. 7.14. Observational drawing of the crater Archimedes. Credit: Peter Grego

can be discerned, projecting westward from the broad black shadow cast by the eastern wall. Archimedes' interior walls are terraced, and outside its sharp rim there are well-rounded flanks indented with a prominent concentric groove. Much of its original external impact structure has been obliterated by Mare Imbrium, although it can be traced in the highlands of **Montes Archimedes** to the south, including a narrow radial crater chain some 50 km long, a section of which can be resolved through a 100 mm telescope. Montes Archimedes sprawl across an area of around 45,000 sq km, and its highest peaks rise to more than 3,000 m.

Montes Spitzbergen, 70 km from the northern wall of Archimedes, are a collection of prominent mountain peaks arrayed over 60 km from north to south, and shaped a little like Svalbard, the terrestrial feature after which they are named. The highest of Spitzbergen's peaks rises to 1,500 m above the surrounding plain. Low angles of illumination will show that this cluster of mountains is actually part of a complex of wrinkle ridges, hills and mountains that marks one of the original walls of the Imbrium multiringed impact basin.

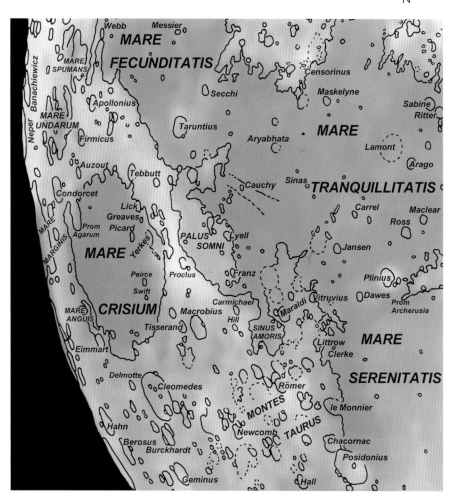

Area Three

Mare Crisium (Sea of Crises) dominates this area of the eastern sector of the
Moon, north of the equator. The eastern part of **Mare Tranquillitatis** and the
northern reaches of **Mare Fecunditatis** (Sea of Fertility) are described, in addition
to the near-limb seas of **Mare Marginis** (Border Sea) and **Mare Undarum** (Sea of
Waves). Features of note include the large crater **Cleomedes**, north of Mare
Crisium, along with the bright-ray crater **Proclus** and **Palus Somni** (Marsh of
Sleep) to the west of Mare Crisium. The area is crossed by the sunrise terminator
between new Moon and 4 days, and the sunset terminator crosses over between full
Moon and 19 days.

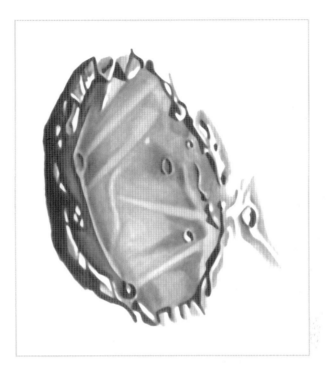

Fig. 7.15. Observational drawing of Mare Crisium. Credit: Peter Grego

Mare Crisium has an oval shape measuring 570 × 450 km, its longest axis oriented east–west. From our terrestrial vantage point, foreshortening causes its east–west axis to be squashed. Its total surface area measures 176,000 sq km. Viewers of the Moon around 2 days after new will be struck by the feature's appearance: bisected by the morning terminator, it makes an imposing dent in the young lunar sickle, visible without optical aid to those with sharp eyes. When Mare Crisium is fully exposed to the morning sunshine a day later, it makes a grand spectacle through any instrument. With its well-defined oval outline, Mare Crisium looks more like a large, flooded crater (which is exactly what it is) than any other mare on the Moon's nearside. Under a high Sun, its surface can be seen to be mottled with darker lava patches, and numerous rays cross the mare, notably those emanating from the bright impact crater **Proclus** (28 km), just beyond its western border. Several large impact craters dot its surface: **Picard** (23 km), **Peirce** (19 km) and **Greaves** (14 km) can be seen under a midday illumination, each having an area of dark albedo lying to its immediate east. Under a low illumination, a concentric system of wrinkle ridges comes into view. These dorsa average about 50 km from the mare border and form a disjointed internal ring. **Dorsum Oppel**, the most prominent of these wrinkles, links with the flooded crater **Yerkes** (36 km) in the west and curves around the northwestern periphery of the mare for 300 km, where it is intercepted by half a dozen narrow wrinkles that cross the mare from the northwestern border. In the northeast can be found the somewhat narrower **Dorsa Tetyaev** (150 km long), and **Dorsa Harker** (200 km) is in the east.

Mare Crisium has imposing lofty mountain borders in the west, whose clean-cut scarp faces shine brilliantly in the morning. As the western mountain border casts broad shadows onto the mare in the early evening, a couple of days after full Moon, the eastern reaches of the sea begin to darken with the encroachment of the terminator, while the mountains of its eastern border shine in the last rays of the setting Sun. A considerable breach exists in the eastern mountain border, where the mare lava has flowed into outlying craters and valleys, notably **Mare Anguis** (Serpent Sea), one of the smallest lunar maria, an irregular dark patch measuring about 200 km from north to south. A large mountainous headland, **Promontorium Agarum**, projects into Mare Crisium from its southeastern shore.

Cleomedes (126 km), an imposing crater, makes a prominent northern neighbor to Mare Crisium. It has a smooth floor on which lie two small craters in the south, a narrow range of hills just north of center and **Rima Cleomedes** in the north, a linear rille 30 km long that branches into two at its eastern end. These can all be discerned through a 100mm telescope. Cleomedes' northwestern wall is intruded upon by several craters, the largest of which, **Tralles** (43 km), is distorted into an ear-shape. To the north lies **Burckhardt** (57 km), an unusual-looking crater with prominent lobes to its northeast and southwest, an example of a larger crater that has been superimposed upon two smaller, preexisting craters. To Mare Crisium's northwest lies the sharp-rimmed, internally terraced crater **Macrobius** (64 km), and **Tisserand** (37 km) to its east. Macrobius adjoins the small, dark-lava patch of **Lacus Bonitatis** (Lake of Good), and further west is **Sinus Amoris** (Bay of Love), a smooth, dark plain 250 km long that forms a northern branch of Mare Tranquillitatis. Some 75 km due west of Mare Crisium's western border, the crater Proclus draws the observer's attention. Proclus is sharp-rimmed and markedly pentagonal in outline, and it is the center of a prominent ray system. Unlike most other rays, Proclus's rays do not spread out in all directions: they are arranged in a broad fan-shape, covering an angle (viewed south at top) from about 5–12 o'clock. The two extremities of the ray system are marked by particularly bright, well-defined rays that each stretch to a distance of about 150 km, and western Mare Crisium itself is liberally dusted with several less distinct rays. None of the ejecta

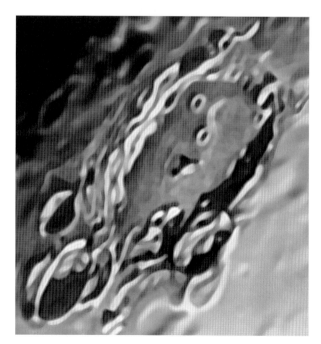

Fig. 7.16. Observational drawing of the crater Cleomedes. Credit: Peter Grego

appears to cover any of the light gray hilly terrain of **Palus Somni**, to the west of Proclus.

Eastern Mare Tranquillitatis is noticeably less wrinkled than the west. Its border with Palus Somni is marked by **Sinus Concordiae** (Bay of Concord). Nearby, the area around the sharp, bowl-shaped crater **Cauchy** (12 km) makes a most interesting area to scrutinize at high magnification when illuminated by a low Sun. **Rima Cauchy** is a prominent linear rille, 210 km long, that cuts northwest–southeast across Mare Tranquillitatis. On the other side of Cauchy, and running roughly parallel to Rima Cauchy, is **Rupes Cauchy**, a fault feature that is part scarp, part rille, 150 km long. To its south can be observed **Cauchy Tau** and **Cauchy Omega**, two sizable round domes visible through a 100mm telescope. A 150mm telescope will reveal the tiny crater on the summit of Omega Cauchy.

A range of low scattered mountains, **Montes Secchi**, separates eastern Mare Tranquillitatis from northwestern **Mare Fecunditatis**. **Taruntius** (56 km) presides over a fairly bland region full of gnarled hills and unimpressive, highly eroded craters like **da Vinci** (38 km), **Lawrence** (24 km) and **Secchi** (25 km). Taruntius itself is an interesting crater, with low walls, an inner ring of eroded hills and a central peak. The crater has a distinct double-walled appearance when it is near the terminator. Despite considerable flooding and erosion, Taruntius' outer flanks have retained much of their original impact sculpting, and radial ridges can be traced for tens of kilometers across northern Mare Fecunditatis.

Mare Marginis straddles the 90°E line of longitude due east of Mare Crisium. Irregular in outline and measuring 360 km from east to west, Mare Marginis has a total surface area of around 62,000 sq km. Lying within the libration zone, it sometimes disappears around the limb completely, but it can usually be glimpsed as a narrow, elongated dark area on the eastern limb during the first half of the lunation, up to full Moon. Other features that may be glimpsed at a favorable libration are the large craters **Hubble** (81 km) and **Goddard** (89 km), along with the giant crater **Neper** (137 km), in north, central and southern Mare Marginis, respectively.

Mare Undarum, southeast of Mare Crisium, lies sufficiently on the Moon's near side to always be on view during the first half of the lunation (with the exception of the first couple of days, when detailed visual observation is impractical in any case). Like Mare Marginis, it is irregular in outline, but it is somewhat smaller, having a diameter of around 200 km and a surface area of around 21,000 sq km. Notable features in its vicinity are the dark, flat-floored crater **Firmicus** (56 km), adjoining the smallest lake on the Moon, **Lacus Perseverantiae** (Lake of Perseverance), just 70 km across.

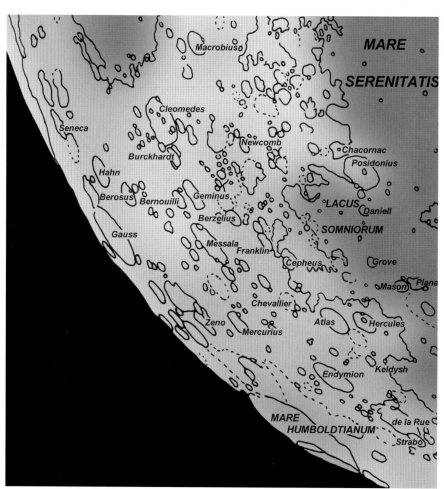

Area Four

The Moon's far-northeastern region, prone to the worst excesses of libration, contains two noteworthy near-limb features, **Mare Humboldtianum** (Humboldt's Sea) and the very large crater **Gauss**. Both of these large formations can be viewed through binoculars at a favorable libration, but both are equally capable of being pushed around the edge of the Moon at an unfavorable libration. Other features of note include the prominent craters **Endymion, Atlas** and **Hercules,** which are most interesting to contrast and compare, **Lacus Temporis** (Lake of Time) and the big craters **Messala** and **Geminus** to their southeast. The sunrise terminator crosses this area between new Moon and 4 days, and sunset takes place from after full Moon to 19 days.

Atlas (87 km) and **Hercules** (67 km) are a prominent duo 30 km apart on the northern border of Lacus Somniorum. Atlas, in the east, has a sharp rim and a substantial inner wall that displays terracing. Through a 150mm telescope, a rough looking floor resolves into a collection of five narrow, sinuous rilles surrounding a collection of low hills. Under a high illumination, Atlas's rim can be clearly discerned, and two very prominent, well-defined dark, circular spots, each about 10 km in diameter, can be seen within. One lies on the northern floor, the other in the south amid the terracing of the inner wall, and they have no obvious topographical associations other than that they each appear to lie at points where two sinuous rilles begin to branch out across the crater floor. A light ray, emanating from the bright crater Thales some 440 km to the northwest (see Area Two), passes between these two dark spots. Adjoining Atlas to its north is an older, much more eroded crater that is only visible under a low illumination. Hercules also has a sharp rim and interior terracing. A prominent 12 km crater, just south of center, and a tiny isolated central hill, occupy an otherwise smooth floor which is darkest to the north. Under a high illumination, Atlas, with its light floor and two dark spots, and Hercules, with its dark floor and single bright spot, make a noteworthy sight. When near the terminator, the intricate impact sculpting of the landscape surrounding both craters can be seen.

Lacus Temporis, to the south and east of Atlas, is an extensive, but patchy, area of fairly smooth undulating terrain, slightly darker than the uplands that surround it, measuring approximately 250 km in diameter with a surface area of about 50,000 sq km. In the hilly plains to its south, halfway to Lacus Somniorum (see Area One), is another prominent crater duo, **Franklin** (56 km) and **Cepheus** (40 km). Both are fairly deep craters with sharp rims and internal terracing. Franklin has a small central mountain and, under a high illumination, the floor surrounding it appears very dark. A bright crater intrudes upon Cepheus's northern rim.

Endymion (125 km), a large and very prominent dark-floored crater, is located to the north of Lacus Temporis, about 200 km northeast of Atlas. Located at longitude 56.5°E, Endymion always appears close to the limb. Endymion has large, somewhat disintegrated internally terraced walls that surround a smooth, dark plain, across which can be discerned linear streaks of light ray material from Thales, 190 km to the northwest. A 150mm telescope will reveal little topographic detail on the crater's smooth floor.

Endymion stands out using binoculars, and since it is always located on the earth-turned lunar hemisphere, it can be used as a guide to viewing **Mare Humboldtianum** on the northeastern limb during the phases leading up to full Moon. Measuring around 200 km from east to west, Mare Humboldtianum is a wide, crescent-shaped dark lava plain that occupies the center of a much larger, 640-km-diameter impact basin whose outer edge passes a little distance east of Endymion. Mare Humboldtianum's eastern edge lies on the 90°E line of longitude, and at times the feature can be librated completely around the other side of the limb. It is, however, visible for most of the time leading up to full Moon, and it makes a rather easy object to discern through binoculars at a favorable libration. A clearly defined mass of light-colored ejecta from the crater **Hayn** (87 km), 135 km to the north, overlies Mare Humboldtianum's northwestern floor, this ray can be traced for almost 1,000 km south to **Geminius**. The huge crater **Belkovich** (200 km), which lies squarely on the 90°E line, intrudes upon Mare Humboldtianum's northeastern border, its northeastern wall encroaching upon the highly disintegrated remnants of the Humboldtianum basin's northeastern wall.

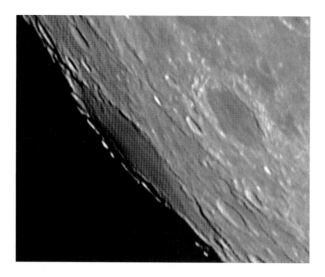

Fig. 7.17. Mare Humboldtianum and Endymion. Credit: Brian Jeffrey

While Mare Humboldtianum itself is easy to see at a favorable libration, the basin's western edge is a somewhat more difficult to discern, requiring a favorable libration between 1 and 2 days after full Moon. If the terrain along the terminator south of Mare Humboldtianum is scrutinized closely, radial impact sculpting from the Humboldtianum basin, both between the dark mare itself and the basin's southern edge, and beyond the basin edge into the landscape beyond, can be seen. This takes the form of distinct lines of craters, valleys and ridges that appear rather eroded. Prominent shadows are cast by several of these ridges, stretching for around 300 km, south of **Zeno** (65 km) and to the north of **Berosus** (74 km), reaching a distance of 700 km from the center of the Humboldtianum basin itself. Few lunar observers have ever recognized the large crater **Compton** (162 km), whose center lies at 115°E, way to the east of Mare Humboldtianum and wholly on the Moon's far side. Compton is one of the furthest placed, clearly recognizable, far side features, and its western rim, seen in profile, is worth looking out for at times of very favorable libration.

When Zeno approaches the evening terminator, a large, disintegrated crater (170 km in diameter) to its south fills with shadow. This is noteworthy because the feature is currently undesignated. Some distance further south lies **Gauss** (177 km), one of the largest craters occupying the Moon's libration zones and subject to occasional librational banishment beyond the limb. Gauss has a low, narrow wall and its floor has a number of smaller craters to the south. Midway between Gauss and Atlas can be found **Messala** (124 km), a large, ancient crater of about the same age and level of erosion as Gauss. Northeast of Messala is the tiny, dark lava patch of **Lacus Spei** (Lake of Hope), 80 km in diameter. To the south of Messala is the prominent, deep crater Geminus (86 km), circular in outline with a sharp rim, terraced internal walls and a small spine of central hills that rises from a flat floor.

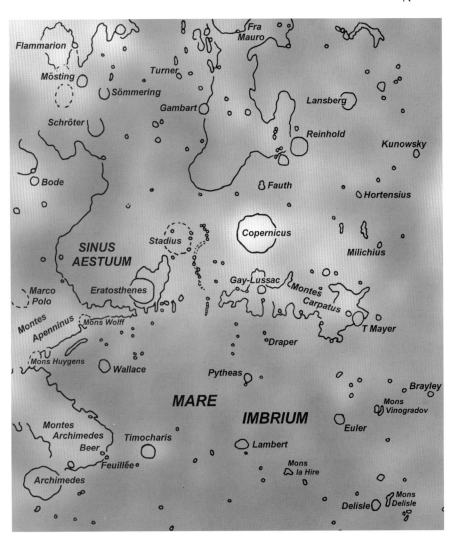

Northwestern Quadrant

Area Five

Mighty **Copernicus**, a large crater with a spectacular system of rays, dominates this area. North of Copernicus, **Montes Carpatus** (Carpathian Mountains) mark part of the southern border of Mare Imbrium, and to their east lies **Eratosthenes,** at the far end of the Apennines. To the southeast is the dark lava plain of **Sinus Aestuum** (Bay

Fig. 7.18. CCD image of Copernicus and Eratosthenes. Credit: Peter Grego

of Billows). Numerous domes and clusters of domes can be found in the general vicinity. Sunrise takes place from 8 to 10 days and sunset from around 23 to 25 days.

Southern Mare Imbrium is relatively smooth and level, dotted here and there with a number of medium-sized craters and mountain peaks, among which roams a system of wrinkle ridges that are generally arranged in a radial fashion to the center of the mare. In the northeast of this area, south of Archimedes (see Area Two), lies the sizable rectangular upland plateau of Montes Archimedes, measuring about 150 km from east to west, which rises from the surrounding plain to a jumbled mass of peaks that rise to heights of 2,000 m. To their west are the little, bowl-shaped twin craters **Feuilée** (9.5 km) and **Beer** (10.2 km), notable for a nearby dome and a tiny crater chain to the east, the latter only visible in a 150mm telescope. **Wallace** (26 km), 310 km from the southern mare border, has a thin, narrow ring representing the rim of a nearly completely submerged crater. Feuilée and Beer lie along an unnamed wrinkle ridge that curves southwards towards the mare border. Another sharp-rimmed bowl crater, **Bancroft** (13.1 km) perches in the mountains immediately west of Archimedes (see Area Two), and to its west a 100mm telescope will show a small linear rille system. **Timocharis** (34 km), 85 km west of Beer, has a sharp polygonal rim, substantial terraced walls and a 6-km-diameter crater in the middle of its floor. **Lambert** (30 km), 180 km to its west, is a similar looking sort of crater, though it is smaller and more eroded than Timocharis. Running southwards across the plain between Timocharis and Lambert, **Dorsa Higazy** meets a ridge extending south of Lambert and proceeds in a strong S shape towards the Apennine border. The ridge is rather prominent under a low morning Sun. South of Lambert, **Lambert R** (50 km) is a ghost ring composed of low wrinkles. A ridge extending from its south links to **Pytheas** (20 km) which itself is linked by a narrow ridge to **Draper** (8.8 km) near the mare border. **Euler** (28 km), a fairly standard-looking impact crater, lies 75 km east of **Mons Vinogradov**, a neat cluster of mountains some 25 km across. A scattered range of smaller peaks lies to its south.

Eratosthenes (58 km), a prominent impact crater, lies at the southwestern end of Montes Apenninus. It has a sharp rim with wide, internally terraced walls and a hilly floor above which rises a group of three individual mountains. Eratosthenes

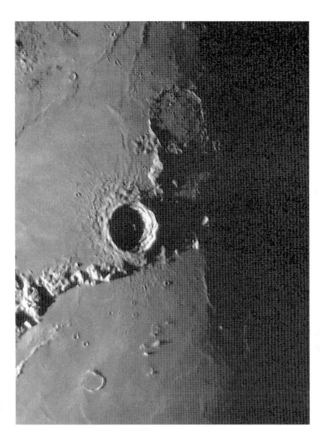

Fig. 7.19. Crater Eratosthenes. Credit: Mike Brown

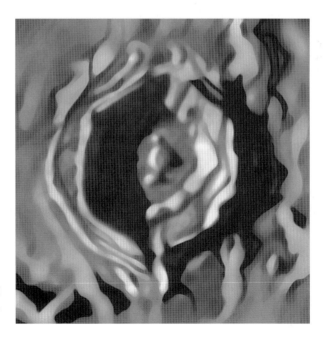

Fig. 7.20. Observational drawing of Eratosthenes. Credit: Peter Grego

displays a considerable degree of external impact sculpting, and radial ridges and secondary impact craters can be traced across the plains of Mare Imbrium to the north and **Sinus Aestuum** to the southeast. A substantial mountain block links the southwestern flanks of Eratosthenes to the northeastern border of **Stadius** (69 km), an unusual flooded crater whose rim is composed of narrow arcs of hills and dotted with tiny craters. Many of the small craters in this vicinity are secondary impact structures from **Copernicus**, 100 km to the west.

Montes Carpatus form part of the southern border of Mare Imbrium. One of the Moon's larger mountain ranges, it measures about 400 km from east to west, with individual peaks that rise to heights of 2,500 m. Like Montes Apenninus to its east, Montes Carpatus displays a strong structuring radial to the Imbrium impact basin. **T Mayer** (33 km) lies at the western end of the range, and **Gay-Lussac** (26 km) in the eastern part, 75 km north of Copernicus. Both craters have sharp-low walls and relatively smooth floors. **Rima Gay-Lussac** (40 km long) lies to the west of Gay-Lussac, and is easily visible with a 100 mm telescope.

Fig. 7.21. CCD image of Copernicus and Montes Carpatus. Credit: Peter Grego

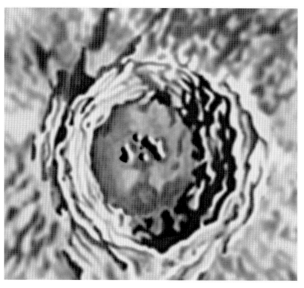

Fig. 7.22. Observational drawing of the mighty crater Copernicus. Credit: Peter Grego

Although it is not the Moon's largest crater, Copernicus (93 km) has an imposing structure and a bright, extensive ray system, which makes it one of the most spectacular craters visible through amateur instruments. Copernicus's ray system spreads in all directions to enormous distances across the lunar surface, and it can easily be seen without optical aid. It is a spectacular sight at full Moon through binoculars or at low magnification through a telescope. Binoculars will resolve Copernicus's rays into an interconnected mass of bright fingers, some of them perfectly linear, others kinked and distorted, that extend to distances up to 800 km. Although the rays look as insubstantial as chalk marks, the piles of ejecta are in places many tens of meters deep.

Copernicus has a generally circular outline with a sharp, rather scalloped rim – the consequence of multiple landslides along weak points in the impact-fractured crust. Inside, its broad inner walls possess multiple levels of intricate terracing. A small crater, Copernicus A (2.5 km) occupies a ledge in the middle of the terracing of the inner eastern wall (discernible in a 150 mm telescope), and there is a prominent kink in Copernicus's eastern rim. A spectacular lighting effect is caused during the early morning and late evening, when Copernicus is partly filled with shadow and some of the higher ridges within the terracing are illuminated by the Sun, amid the shadows. Even a cursory telescopic view will convince the observer that Copernicus's floor is considerably depressed beneath the mean level of the surrounding landscape. Measured from the highest points on the crater's rim to the mean level of the floor, Copernicus is 3,760 m deep. The southern half of the crater's floor is hillier than the north. A 60 mm telescope will not resolve this terrain, and the southern floor appears as smooth as the north, but slightly darker. A group of central mountains, with peaks as high as 1,200 m, rises above Copernicus' floor.

Over a distance of more than 50 km, a complex of concentric hills interlocks with radial ridges and furrows on the crater's outer flanks, rising to a height of 900 m at the rim. Copernicus is a young crater, less than a billion years old, and the fact that it was blasted out of a relatively flat part of the Moon's surface, rather than a mountainous or heavily cratered region, makes the impact sculpting and ejecta system clearly discernible. To the west of the crater rim there is a clear boundary between the concentric ridges and the radial features, forming a broad ledge some 20 km wide. Ridges and furrows, substantial piles of ejected debris mixed with secondary impact craters and crater chains spread in all directions further out from the crater. The ray system spreads much further afield. **Fauth** (12.1 km) and **Fauth A** (9.6 km) are conjoined craters resembling a keyhole, and their orientation suggests that they are most likely secondary Copernican impact structures. Notable crater chains – not all perfectly straight or necessarily radial to the center of Copernicus – can be seen in the surrounding country under a low angle of illumination. The most prominent crater chain lies between Copernicus and Eratosthenes, concentric to Copernicus, from Mare Imbrium to the western wall of Stadius, over a distance of 200 km. It is made up of an interlocking chain of small craters that range from around 1 to 7 km in width. A radial chain of more elongated craters runs for 80 km northwest of Copernicus. At first sight, Rima Gay-Lussac (see above) also appears to be part of Copernicus' external impact structuring, but it is in fact an older, more eroded feature, perhaps a sinuous rille cut in the lava plains south of Montes Carpatus a couple of billion years before Copernicus was formed.

Mare Insularum (Sea of Isles) is a broad, patchy sea to the south of Copernicus that stretches for a distance of about 900 km from **Kepler** in the west (see Area

Seven), to Sinus Aestuum and Sinus Medii in the east (see Area One), where there are several extensive dark albedo markings that are easily visible through binoculars around full Moon. One of these dark markings covers an area of about 5,000 sq km in the hills to the east of Mare Insularum. To its south are the disintegrated craters **Schröter** (35 km) and **Sömmering** (28 km), both of which have flooded floors and breached southern walls. **Mösting** (26 km), a short distance to the southeast, is a prominent, sharp-rimmed crater with broad, internally terraced walls. Its northern rim is interrupted by a ridge that cuts across the crater's inner wall. The sharp-rimmed, flat-floored crater **Gambart** (25 km), is 280 km to the west. Midway between Gambart and Schröter lie two sharp-bowl craters, **Gambart C** (12.2 km) and **Gambart B** (11.5 km). Under a low illumination, a large dome with a base diameter of 18 km can be seen to the immediate southwest of Gambart C. West of Gambart, up to **Reinhold** (48 km), a scrambled mass of hills rises from the patchy mare. Reinhold is a prominent crater with a strongly terraced interior wall and a

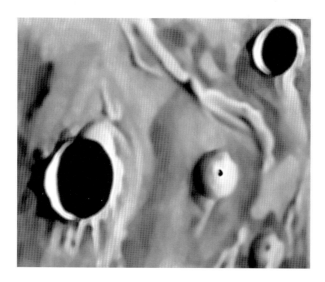

Fig. 7.23. Observational drawing of crater Milichius and the small dome Milichius Pi. Credit: Peter Grego

Fig. 7.24. CCD image of crater Milichius and the small dome Milichius Pi. Credit: Peter Grego

smooth floor containing a few small hills. A magnificent collection of sizable domes can be found in the area of Mare Insularum south of Montes Carpatus, west of Copernicus. Copernicus is so overwhelmingly spectacular that the observer is tempted to linger over it for hours on end, but it is worthwhile paying closer scrutiny to these least obvious attractions of the region. South of T Mayer, jostling for position among a number of scattered peaks, is an extensive field of at least a dozen immediately discernible domes, which on average measure about 12 km across at their bases. Several of these have small craters at their summits, which can be resolved in a 150mm telescope, but the group itself is a little challenging to discern at low angles of illumination because of the shadows cast by mountains in the vicinity. Across the mare to the south is the bowl crater **Milichius** (13 km) and the small, but easily observable, isolated dome **Milichius Pi** (10 km across). Between Milichius Pi and the dome field to the north can be seen a very large, elongated domelike swelling, some 50 km wide, its southeastern slopes punctured by a cluster of small hills. At 115 km southeast of Milichius lies another prominent bowl-shaped crater, **Hortensius** (15 km), and to its north lies a cluster of six circular domes, averaging 10 km in diameter, most of which have summit craters that can be resolved in a 150 mm telescope.

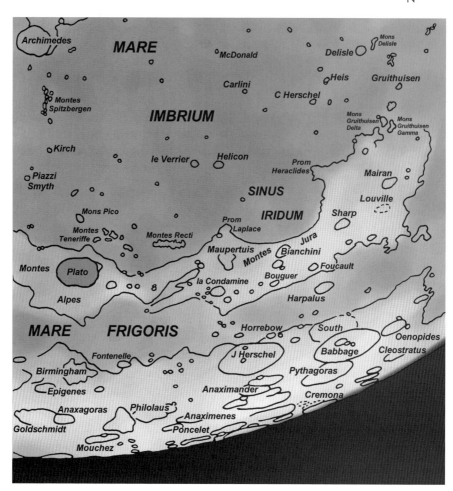

Area Six

Some prominent mountains poke out of Mare Imbrium near its northern shoreline. They include **Montes Recti** (Straight Range) and **Montes Teneriffe**. Northwestern Mare Imbrium is dented by the large bay **Sinus Iridum** (Bay of Rainbows) and bordered by the splendid mountain arc of **Montes Jura**. The magnificent, dark-floored crater Plato is sunk into the western heights of the lunar Alps. The dark plains of **Sinus Roris** (Bay of Dew) flow northwest of Mare Imbrium's mountain border to join with the western reaches of Mare Frigoris. The cratered highlands further north are home to the large craters **J Herschel**, **Babbage** and **Pythagoras**. Sunrise takes place between around 7 and 11 days; sunset between around 21 and 25 days.

With a surface area of 3,554,000 sq km, **Mare Imbrium** is the Moon's largest circular mare, second only to Oceanus Procellarum in area. Mare Imbrium has a diameter of 1,160 km, and is bordered by prominent mountain ranges to its north, east and south, but its western reaches blend into eastern Oceanus Procellarum. Mare Imbrium occupies an inner ring of the much larger Imbrium multiringed impact basin, the outer rings of which have diameters of 1,700 and 2,250 km, and perhaps as great as 3,200 km, encompassing Oceanus Procellarum itself.

Mare Frigoris runs parallel to the northern border of Mare Imbrium, separated from it by a substantial strip of mountains, from the Montes Jura in the west to the Montes Alpes in the east. Parts of the midwestern section of Mare Frigoris, north of Plato, display a little more wrinkling than the rest of the sea. Mare Frigoris blends into the plains of **Sinus Roris** around the crater **Harpalus** (39 km), a prominent crater with a sharp polygonal rim and the center of a small, somewhat faded system of rays. In the same area, rays from Anaxagoras, several hundred kilometers to the northeast, can be clearly traced. North of Harpalus, on the shore of Mare Frigoris, the crater **South** (108 km) has a low, eroded wall that appears squared-off in the east. South joins with the wall of **Babbage** (144 km), another ancient and eroded crater, its hilly southern floor dented by a younger impact crater, **Babbage A** (26 km). **Oenopides** (67 km) is sunk into the highlands west of Babbage.

Joining Babbage's northeastern wall is the imposing crater **Pythagoras** (130 km). Pythagoras is slightly larger than Copernicus (see Area Five), but they are remarkably similar-looking craters when viewed from above. Like Copernicus, Pythagoras has a scalloped rim, broad walls with intricate layers of terracing and a hilly floor (hillier in the south), from which protrudes a group of central mountains. Because the crater has been carved from highland crust, the external impact structure so clearly seen in the vicinity of Copernicus is generally absent from the surroundings of Pythagoras. Pythagoras lies just outside the libration zone, and although it never disappears completely beyond the northwestern limb, an unfavorable libration will make it appear extremely foreshortened. At a favorable libration and angle of illumination, observers can gain at least some idea of how grand this crater really is, as the terraced northwestern wall shines beyond central mountains that are seen in profile against each other. This is how the interior of Copernicus must appear when viewed from a low angle. Beyond Pythagoras and into the northwestern libration zone lie the craters **Boole** (63 km) and **Cremona** (85 km), which can only be observed at a favorable libration a day or so prior to full Moon.

J Herschel (156 km) lies further east along the northern shore of Mare Frigoris. Although it is one of the near side's largest craters, J Herschel is rather eroded, with a battered wall and a hilly floor with a couple of small craters, but it is very prominent when near the terminator. Far-western parts of its floor are stained with dark albedo markings, and a 200 mm telescope will reveal several small rilles in this area. To the north of J Herschel can be found a linked group of large, flooded craters, **Anaximander** (68 km), **Anaximander B** and **Anaximander D**, intruded upon in the east by the prominent crater **Carpenter** (60 km). **Pascal** (106 km) and **Brianchon** (145 km) are two large craters that lie in the libration zone beyond Carpenter. A large part of the area east of Carpenter is relatively smooth and covered with light rays and ejecta from Anaxagoras. **Philolaus** (71 km), a deep crater with terraced walls and sizable peaks rising from its floor, is the most prominent landmark in this area. Low illumination will reveal that Philolaus is superimposed upon a larger, more ancient and eroded crater to its southwest. At the same time, a highly eroded, disintegrated ancient crater (unnamed, about 95 km across)

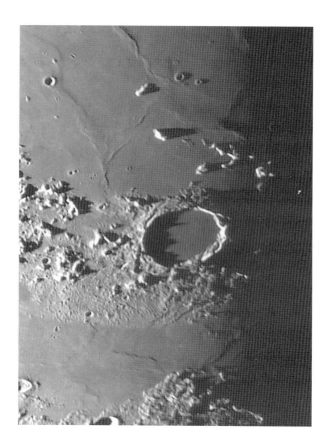

Fig. 7.25. Plato. Credit:
Mike Brown

can be seen immediately southeast of Philolaus. Further towards the limb can be found the relatively smooth-floored crater **Anaximenes** (80 km) and **Poncelet** (69 km) to its north, the latter lying in the northern libration zone.

On the northern coast of Mare Frigoris, 310 km northwest of Plato, is the clean-cut circular crater **Fontenelle** (38 km), notable for its small near-central craterlet. From its southern wall, wrinkle ridges run south across the mare to join with several others, to meeting with the highlands to the north of Plato. Plato (101 km) is a majestic crater, one of the most prominent landmarks in this sector of the Moon. Circular in outline, with a somewhat dented rim, Plato's flat, dark floor is sunk more than 2,000 m below the level of the surrounding terrain. There are no traces of the crater's original central elevations, as they have been completely submerged by lava flows. The only topographic detail visible on Plato's floor is five small craters, ranging from 1.7 to 2.2 km in diameter. Under a low Sun, these craters are easy to spot in a 100mm telescope, since their raised rims shine brightly and they cast noticeable shadows across Plato's floor. Under a high Sun, however, they appear as minute bright dots that can be challenging to discern through a 100mm telescope. Around the edge of the floor can be resolved piles of material that have slumped from the inner wall. A large, triangular section of the inner western wall – a block with a surface area of about 50 sq km – has broken away and slipped towards the floor, leaving a sizable dent in the crater's rim. There are no overt indications of impact sculpting around Plato of the kind that surrounds similar-sized craters. A short distance west of Plato, the prominent crater **Plato A** (23 km) is sunk deep into the mountains.

Illuminated by an early morning or late evening Sun, the shadows cast by Plato's walls onto its floor are fascinating to observe. In the morning, while the western flanks of Plato remain in shadow, joined with the terminator, its inner western wall and the western part of its floor are illuminated by shafts of sunlight streaming through low points in the crater's eastern rim. As the shadow cast by the eastern rim recedes, the edge of the shadow projects into four or five long, pointed fingers, which shorten rapidly as the Sun rises. In the evening, as Plato's eastern flanks have begun to be surrounded by the darkness of the terminator, its inner eastern wall gleams as a bright crescent in the rays of the setting Sun as the floor darkens. One particularly long shadow, cast by a high part of the rim (to the north of the major landslide mentioned above), touches the base of the eastern wall. Several more long shadow fingers soon project completely across the floor, and, within a matter of hours, the whole of Plato's interior, except the inner eastern wall, is plunged into darkness. The appearance and orientation of the shadows within Plato are never the same from one lunation to the next because of the effects of libration and the change in the direction of illumination by the Sun that it causes.

Shadowplay is also fascinating to observe in the area of Mare Imbrium, to the south of Plato. **Montes Teneriffe** is a scattered range of mountains with a length of around 110 km, comprising a Y-shaped spine of mountains with a substantial block in the west and a smaller one to the east. Its highest individual peaks rise to 2,000 m. Using a 150 mm telescope at high magnification, there are indications of small craters on the summits of the Montes Teneriffe when they are illuminated by a low Sun. A short distance southeast of Montes Teneriffe, the hefty mountain block of **Mons Pico** (base size 15 × 25 km) rises to a height of 2,400 m above the marial plain. Under a midday Sun, Mons Pico is by no means uniformly bright – much of its northern heights are taken up by a distinct, dark oval area bordered by brighter strips, with a sizable bright area in the far north. Illuminated by a low morning or evening Sun, Mons Pico casts a prominent broad, long shadow onto the mare. As the evening terminator engulfs the mountain, it breaks up into a number of bright, individual components that continue to shine for some hours as the darkness rises inexorably from the base of the mountain.

Under a low illumination, Montes Tenerife and Mons Pico appear to mark part of the southern rim of a largely buried crater that occupies the mare up to its mountain border, south of Plato. Unofficially known as **Ancient Newton** (115 km), traces of its buried eastern wall can be discerned in the wrinkle ridges that run from Mons Pico to the east. However, it is possible that the arrangement of wrinkle ridges does not reflect the underlying topography, and that a buried crater does not really exist here. Wrinkle ridges run further south across the mare from Mons Pico, past the elongated mountain **Mons Pico Beta** (9 × 20 km) to loosely link up with wrinkle ridges that curve past Montes Spitzbergen (see Area Two) and around to the west in the **Dorsum Grabau**, parallel to the southern border of Mare Imbrium. Several other wrinkle ridges proceed south from Montes Tenerife, fading out in mid-mare.

Just 50 km from the northern shoreline of Mare Imbrium, to the west of Montes Tenerife, the long, straight massif of **Montes Recti** is one of the Moon's most remarkable -looking mountain ranges. Oriented precisely east–west, this 78-km-long bar of mountains averages 20 km wide, and has at least 20 individual peaks, the highest of which rise to 1,800 m. Looking much like a large, segmented centipede, the range displays a structure that reflects the radial impact sculpting of the Imbrium basin. A sizable crater, 8 km wide, dents the eastern end of the range. To the west of Montes Recti, the coastline of Mare Imbrium projects southwards at **Promontorium Laplace**, a prominent headland that marks the eastern border of **Sinus Iridum**, a well-defined bay in Mare Imbrium measuring 260 km in diameter. Sinus Iridum is an impact basin that has been flooded with lava, completely submerging half its wall. Traces of buried structure can be seen in a number of low wrinkle ridges on the border of Sinus Iridum and Mare Imbrium; they connect Promontorium Laplace with the east-pointing headland of **Promontorium Heraclides**, on the opposite side of the bay. The ridges curve for a considerable distance south of Promontorium Heraclides, joining the sharp-rimmed crater **C Herschel** (13 km) with **Dorsum Heim**, which runs a further 130 km to the south.

Southeast of Sinus Iridum, the craters **Helicon** (25 km) and **le Verrier** (20 km) are the most prominent landmarks in this section of mare Imbrium. Both have sharp rims, traces of internal terracing and slightly bumpy floors. Helicon lies along a broad but low ridge, visible only at low angles of illumination. Each is surrounded by a fairly homogenous mantle of light (but not prominent) ejecta, and both lie within an extensive, but subtle patch of slightly darker albedo, measuring about 280 km across.

Sinus Iridum's mountain border is made up of the prominent arc of **Montes Jura**, which rises in a series of disjointed terraces from the shoreline to heights of more

Fig. 7.26. Montes Jura and Sinus Iridum. Credit: Mike Brown

than 4,000 m. **Bianchini** (38 km), a deep, clear-cut crater, lies in the mountains immediately north of Sinus Iridum. The crater **Maupertuis** (46 km) lies 100 km to its east, considerably eroded and displaying structure radial to Sinus Iridum. A 150mm telescope will reveal the sinuous rilles of **Rimae Maupertuis** to its east, which wind their way across the mountainous cratered terrain.

West of Sinus Iridum, the prominent crater **Sharp** (40 km) has broad, terraced inner walls and a central elevation. **Mairan** (40 km), a similar looking crater but with a flat floor, lies among the bright, crater-peppered (unnamed) uplands further west of Sinus Iridum. One of the Moon's most eroded named craters, the extremely battered **Louville** (36 km) can be just made out on the shore of a small bay in Sinus Roris. To its west, a very long, though exceedingly fine, sinuous rille, **Rima Sharp**, makes its way for more than 200 km over the dark plains of eastern Sinus Roris. Northern parts of this rille are visible using a 150 mm telescope on nights of excellent visibility. Another narrow, sinuous rille, **Rima Mairan**, runs 100 km further south, closer to the shoreline.

Towering over the southern tip of the uplands south of Mairan are two huge, rounded mountains, **Mons Gruithuisen Gamma** and **Mons Gruithuisen Delta**. Mons Gruithuisen Gamma is domelike, almost round in outline with a southern spur, and has a base measuring 20 km across. A 100mm telescope will resolve a small crater on its summit. A narrow valley separates it from Mons Gruithuisen Delta to the east; this is a rectangular mountain mass with rounded slopes and a base measuring some 25 km. To the south of this prominent pair of mountains, Mare Imbrium's northwestern border is marked by numerous peaks, hills and wrinkle ridges. The small crater **Gruithuisen** (16 km) lies at the northern end of **Dorsum Bucher** (90 km long), one of a series of ridges in the area that run parallel to the border of Mare Imbrium. Two interesting (unnamed) small knobbly plateaus rise from the mare, one to the north of Gruithuisen, the other to the south. Further southeast, the craters **Delisle** (25 km) and **Diophantus** (19 km) are near neighbors and make a prominent landmark, along with the arrowhead-shaped **Mons Delisle** (30 km long) to the west, joined to Diophantus' rim by a prominent ridge.

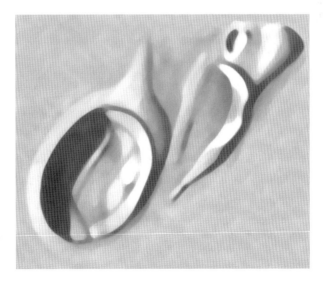

Fig. 7.27. Observational drawing of crater Delisle and Mons Delisle. Credit: Peter Grego

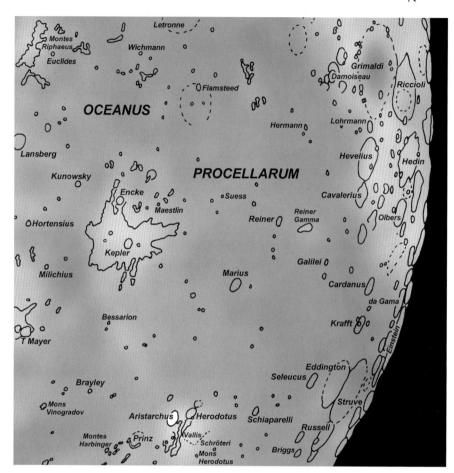

Area Seven

Oceanus Procellarum (Ocean of Storms) covers a major portion of the western part of the Moon's near side, almost touching the western limb north of the equator. The bright **Aristarchus Plateau** in the north is of exceptional geological interest. The brilliant crater **Aristarchus** and its bright ray system dominate the north of the area. Aristarchus contrasts with its neighbor, the dark-floored crater **Herodotus**. Nearby is the Moon's largest sinuous rille, **Vallis Schröteri** (Schröter's Valley), and many smaller sinuous rilles can be observed in the vicinity. **Kepler** and its rays dominate the southeastern part of the area. At the center of Oceanus Procellarum, **Marius** lies on the edge of a vast field of domes. To its southwest is the enigmatic bright swirl, **Reiner Gamma**. Near the western limb are the prominent large craters **Hevelius** and **Hedin**, and further north along the limb are the linked

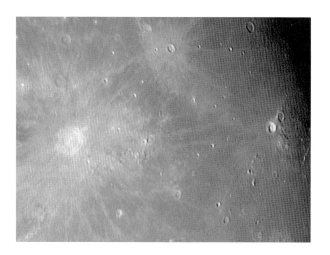

Fig. 7.28. Prominent rays from the bright craters Copernicus, Kepler and Aristarchus spread across the Moon's western maria. Credit: Brian Jeffrey

trio of the large plains, **Eddington**, **Struve** and **Russell**. Sunrise takes place from around 10 days to full Moon. Sunset is from around 25 days to new Moon.

Aristarchus (40 km) may be on the small side, but its brilliance and commanding position in northeastern Oceanus Procellarum make it one of the Moon's most prominent craters. It has a very sharp rim and a somewhat polygonal outline, with strongly terraced inner walls that, at certain illuminations, give it an appearance rather like a Roman amphitheater. While the inner western wall is presented very favorably, our view from an angle of about 42° above the crater means that (when it is illuminated by the Sun) its inner eastern wall appears as a narrow sliver. This gives the impression that Aristarchus has an offset floor with differently sloped inner walls. In reality, its inner walls maintain the same breadth around the crater and its floor is perfectly central. Parts of Aristarchus's inner and outer western

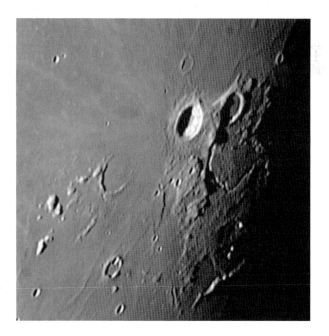

Fig. 7.29. The brilliant crater Aristarchus. Credit: Brian Jeffrey

walls, and its central hill, are among the most reflective surfaces on the entire Moon – they can appear dazzlingly bright under a high Sun. Several dusky bands of lower albedo mark the inner western wall; these shadow bands are a source of fascination to many lunar observers. They can easily be observed through a 60 mm telescope whenever the inner western wall is illuminated. Aristarchus's rather small floor measures about 19 km across, and lies about 3,000 m beneath the rim. At its center can be seen a small central hill. The crater's rim rises several hundred meters above the surrounding terrain, with one major terrace halfway up the slope. At a low morning illumination, when the crater is emerging from the terminator, the outer terrace shows up as a bright arc separated from the bright rim by a line of shadow, surrounded by a mass of radial impact sculpting further afield.

Aristarchus has a prominent ray system – a keen-sighted person can even discern the brightening in the Aristarchus region without any optical aid. Through binoculars, the rays can be traced for hundreds of kilometers to the south and east, but they are not as bright as those of Copernicus (see Area Five) or Kepler, with which Aristarchus's rays mingle. Surrounding Aristarchus (with the exception of its western flanks) is a dark collar extending about 25 km from the crater's rim, representing an area that has been smothered by dark, glassy impact melt. There also appears to be an absence of rays across the Aristarchus Plateau to the northwest, in an area that is often called (unofficially) "Wood's Spot", which appears distinctly orange through the telescope eyepiece.

Herodotus (35 km), a short distance west of Aristarchus, is an altogether less showy sort of crater, with low, unterraced walls and a smooth, flat floor. A brilliant lobe of ejecta from southwestern Aristarchus crosses to the south of Herodotus, coating the southern part of its dark lava-flooded floor with a bright ray. The crater's northern wall is breached by a broad, roughly hewn valley some 18 km long, at the northern end of which sits a prominent 7 km bowl-shaped crater. This crater sits at the head of **Vallis Schröteri**, the Moon's largest sinuous rille. In a formation unofficially known as the Cobra's Head, the valley widens out to about 10 km, and narrows again further south, where it maintains an average width of around 5 km along much of its 160 km length. Vallis Schröteri cuts across the

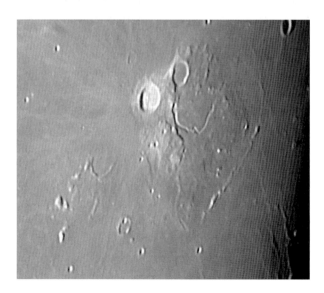

Fig. 7.30. CCD image of the Aristarchus plateau, with the brilliant Aristarchus, Herodotus and Vallis Schröteri. Credit: Peter Grego

Aristarchus Plateau in a contorted, generally semicircular, path, gradually narrowing and becoming shallower as it heads southwest, petering out some 75 km northwest of Herodotus. Under a high illumination, the floor of the valley remains visible as a light-colored line. A much narrower sinuous rille meanders down the center of Vallis Schröteri; under excellent conditions, using a 200 mm telescope, it is possible to glimpse parts of this tiny rille in the widest part of the valley.

Vallis Schröteri, cut by the erosive action of fast-moving lava flows, is one of the more obvious signs of volcanic activity in this region of the Moon. A subtler example is the small dome, **Herodotus Omega**, which lies between Herodotus and the north bend of Vallis Schröteri. The northwestern boundary of the Aristarchus Plateau rises from the Oceanus Procellarum near the little crater **Raman** (11 km), and includes the small, isolated peak **Mons Herodotus**. Further north, **Montes Agricola**, a prominent linear mountain ridge, extends for 160 km from southwest to northeast, linked to the Aristarchus Plateau by the umbilical cord of the Moon's smallest named wrinkle ridge, **Dorsum Niggli** (40 km long). **Dorsa Burnet** (200 km long) cross southwards over Oceanus Procellarum to the west of Montes Agricola and terminate at the crater **Schiaparelli** (24 km). Schiaparelli lies near the northern tip of one of two prominent rays that are traceable west of the Aristarchus Plateau across the mare to the bright near-limb crater **Olbers A** (see below).

North of Aristarchus, the plateau abruptly rises yet higher, and becomes slightly more rugged in the vicinity of the small crater Väisälä (8 km). A prominent scarp, **Rupes Toscanelli** (70 km long) cuts cleanly north across the plateau to **Toscanelli**

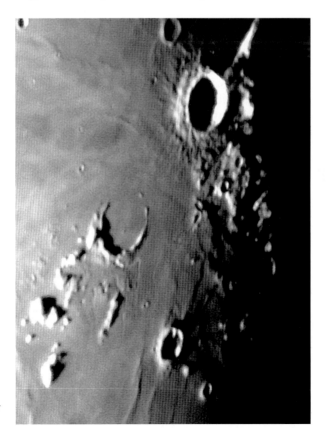

Fig. 7.31. The ghost crater Prinz. Credit: Mike Brown

(7 km). A 150mm telescope will resolve several sinuous rilles, **Rimae Aristarchus**, in this region; many of these begin in small, well-defined craterlets, such as Vallis Schröteri, and flow downhill. Most of these sinuous rilles are in excess of 50 km long. Further east, **Rimae Prinz** is another collection of narrow, sinuous rilles, the longest of which commences in a tiny crater, **Vera** (4.9 km), just north of **Prinz** (47 km), and meanders 80 km to the north. Prinz itself is the remains of a largely submerged crater whose northern rim just protrudes above the level of the mare. **Montes Harbinger**, to its northeast, is a collection of three large mountain blocks, themselves the protruding remnants of part of the submerged western wall of one of the Imbrium basin's rings; they are surrounded by **Dorsa Argand**, a collection of wrinkle ridges that link with **Dorsum Arduino (110 km long)** further south. The southern branch of Dorsum Arduino is cut across by **Rima Brayley**, an exceedingly fine sinuous rille some 240 km long that snakes across the plains of western Mare Imbrium, passing just northwest of the small, sharp-rimmed crater **Brayley** (14.5 km). Rima Brayley can only be observed through a large telescope in excellent conditions.

Due west of Aristarchus, a trio of very large flooded craters, **Eddington** (125 km), **Struve** (170 km) and **Russell** (103 km), dominate the near-limb region of the far western reaches of Oceanus Procellarum. Eddington's substantial northern wall is broad and projects northwards, but its southern rim is largely submerged, traceable only by a few small elevations. Eddington can clearly be seen under a high angle of illumination, and close scrutiny will reveal faint rays from Olbers A running northwards across the crater's floor. To Eddington's east is the prominent, sharp-rimmed crater **Seleucus** (43 km), its eastern flanks crossed by a prominent bright ray emanating from Olbers A. Seleucus's broad, terraced inner walls show up brightly under a high Sun. Eddington shares its low western wall with the broad low-walled plain of Struve, which is linked with Russell to the north by a broad breach in its northern wall. The far western border of Oceanus Procellarum, including Russell's far western rim, lies in the libration zone, so the craters can appear extremely foreshortened, squeezed right up to the limb. At a favorable libration illuminated by the morning Sun, prior to full Moon, it is spectacular to see the western parts of the floors of Struve and Russell remain in darkness while their western rims are illuminated by the rising Sun.

Across the mare due south of Eddington, **Krafft** (51 km) is joined with **Cardanus** (50 km) by **Catena Krafft** (60 km long), a small continuous crater chain that can be discerned through a 100 mm telescope. Cardanus is smothered with ray material from Olbers A, 190 km to the south, and it is among this ray material that **Rima Cardanus** can be found; this is a narrow rille that cuts 120 km from the uplands and across the mare southeast of Cardanus. A 150 mm telescope will resolve this feature. Krafft and Cardanus make a prominent landmark that can be used to hop across to the limb to identify a group of large features that occupy the western libration zone. These features, visible during favorable librations prior to full Moon, include **Balboa** (70 km), **Dalton** (61 km), **Vasco da Gama** (96 km), **Bohr** (71 km) and **Vallis Bohr** (180 km long), the last a large gouge radial to **Mare Orientale** (see Area Eleven). A favorable libration and illumination will also reveal the vast crater **Einstein** (170 km), much of whose western floor lies beyond the 90°W line of longitude. A large, prominent crater, **Einstein A** (53 km) lies almost central on Einstein's floor. Adjoining Einstein's northern wall, on the line of 90°W, is the crater **Moseley** (93 km), north of which lie the prominent flooded craters **Bartels** (55 km) and **Voskresenskiy** (50 km).

Fig. 7.32. Kepler, a prominent ray crater in Oceanus Procellarum. Credit: Mike Brown

South of the Aristarchus Plateau, the area bounded by Seleucus in the west, the Marius hills in the southwest and Kepler in the south is fairly flat, splashed with rays and dotted with small craters. A number of fine, low ridges can be seen among the rays radiating south of Aristarchus – piles of ejecta thrown out by the impact. These merge with the brighter rays of Kepler (32 km), which are so bright that they

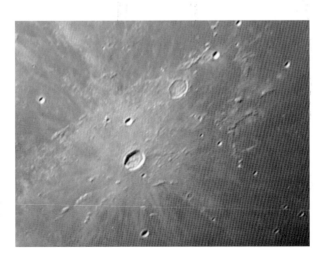

Fig. 7.33. Kepler's fine rays stretch in all directions across Oceanus Procellarum. Credit: Brian Jeffrey

can be seen without optical aid; this forms the southwestern apex of a triangle with Aristarchus and Copernicus. Although Kepler is a prominent, sharp-rimmed crater, close examination will reveal it to be much less grand-looking than, say, Aristarchus. Its walls are much lower and less broad than those of Aristarchus, with only a trace of terracing, and its knobbly floor is not as deep. The crater has impacted on the eastern side of a small, rough plateau, and only its eastern outer flanks display extensive impact sculpting. Kepler's ray system stretches in all directions, some individual rays reaching distances of 600 km or more. A fascinating "swirling" structure can be discerned in the scattered lines of mountains surrounding Kepler; these are best viewed around 6 hours after Kepler has emerged from the morning terminator. Of particular note is a sizable dome with a 10-km-diameter base that lies just northwest of Kepler. Midway between Kepler and Milichius (see Area Five), a very narrow, sinuous rille, **Rima Milichius** (110 km long) makes its way north across Mare Insularum from its origin near a small cluster of mountains. This feature can only be resolved in large instruments. A scattered line of peaks makes its way southeast, across to **Lansberg** (see Area Nine), dividing Mare Insularum in the east from Oceanus Procellarum in the west. The flooded crater **Kunowsky** (18 km) lies along this range.

Encke (29 km), 90 km south of Kepler, is a similar looking crater, though its walls are lower and somewhat more polygonal, and its floor slightly rougher, than Kepler's. Little impact structure can be discerned around Encke. The crater lies on the northeastern floor of a large, flooded crater, **Encke T** (110 km). To the west lie the disintegrated rims of several submerged craters, among which lie **Maestlin** (7 km) and **Rimae Maestlin**, a small group of parallel linear rilles that cut southeast from **Maestlin R**.

Marius (41 km), in the center of this area, is a sharp-rimmed crater with a smooth, flat floor. Sweeping to its south is a system of wrinkle ridges that extends all the way to the southern border of Oceanus Procellarum across a distance of some 700 km. Marius lies on the eastern edge of the **Marius Hills** (the area has no official name), a mass of domes and low hills that covers a fairly well-defined rectangular area of more than 30,000 sq km. When the area is illuminated by a rising Sun, a 100mm telescope at a high magnification will reveal at least 100

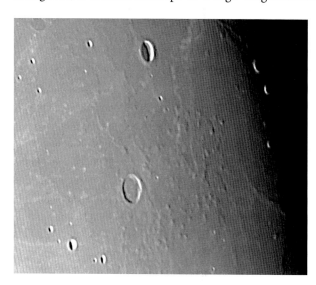

Fig. 7.34. CCD image of Marius and the extensive dome field of the Marius hills in Oceanus Procellarum. Credit: Peter Grego

rounded hills and domes, among which wind several wrinkle ridges. Under a high angle of illumination, the area looks rather patchy and blends into the surrounding mare. A sinuous rille, **Rima Marius** (250 km long), visible through a 150 mm telescope, originates some distance north of Marius and winds around the northeastern border of the Marius Hills. Another sinuous rille, **Rima Galilei** (180 km long), skirts the southwestern border of the Marius Hills.

Reiner (30 km), a prominent, sharp-rimmed crater with a central peak, lies on a wrinkle ridge south of the Marius Hills. Lying along another wrinkle ridge 100 km west of Reiner, the bright patch of **Reiner Gamma** is one of the area's most prominent landmarks. Composed entirely of bright ray material, Reiner Gamma is an elongated oval splash, the core of which extends 60 km east–west. A lobe of bright material proceeds from its northeastern border in a prominent, though disjointed, line some 150 km north to the edge of the Marius Hills. Unlike many ray systems, Reiner Gamma is prominent even when illuminated by a low Sun. It is the near side's best example of a swirl (see Chapter One). The fact that few small craters overlie Reiner Gamma means that it is a comparatively young lunar feature. It may be an area where volcanic gases have vented, causing a change in the coloration of the surface materials. Alternatively, it may have been caused by cometary impact, the area's strong magnetism having been transplanted to the lunar surface from the comet's nucleus.

On the western shore of Oceanus Procellarum, 130 km west of Reiner Gamma, can be found **Planitia Descensus** (Plain of Descent), an unassuming patch of mare that marks the landing site of Luna 9, the first soft-landing lunar probe, in February 1966. To its south and west lies a group of very large craters. **Cavalerius** (58 km), a deep, sharp-rimmed crater with prominent internal terracing, is the center of a faded ray system that can be traced in Oceanus Procellarum. Cavalerius adjoins the northern wall of **Hevelius** (106 km), a prominent crater with eroded walls and a floor that is crossed by numerous linear rilles. Southeast of the crater's near-central peak, the most prominent of **Rimae Hevelius** are arranged in a large X shape, visible in a 100mm telescope. Both rilles proceed to cut through Hevelius' walls and break out into the surrounding terrain. One heads southeast across the mare to about 50 km from Hevelius's wall, east of the crater **Lohrmann** (31 km); the other cuts across about the same distance through somewhat hillier terrain west of Lohrmann. A number of other small linear valleys can be found in the vicinity. Due west of Hevelius, near the western limb, the large, highly eroded crater **Hedin** (143 km) also displays a system of linear rilles on its floor. A patch of dark material on Hedin's northwestern floor is clearly visible under a high illumination, though the rest of the feature is impossible to make out except under a low illumination. Clearly traceable across the walls and floor of Hedin, along with those of many other features in the region, are the sculpting effects produced during the asteroidal impact that created the Orientale multiringed impact basin (see Area Eleven). This is particularly noticeable at a favorable libration combined with a low morning or evening illumination. Due north of Hedin lie **Olbers** (75 km), a somewhat eroded crater, and **Olbers A** (45 km), a bright, fresh young impact crater, the center of a prominent system of rays. A number of these rays stretch vast distances across Oceanus Procellarum; particularly noteworthy is a duo of rays that reach across to the Aristarchus Plateau, almost 900 km to the northeast.

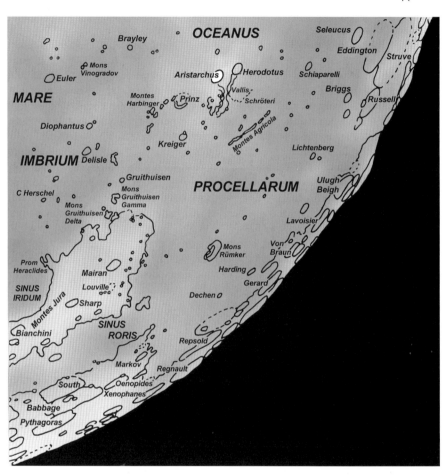

Area Eight

Taken up largely by the smooth plains of the northwestern part of Oceanus Procellarum, this area is notable for **Mons Rümker**, a large plateau made up of a collection of domelike swellings. A few large, low-walled craters lie near the limb, on the western fringes of the mare, notably **Ulugh Beigh**, **Von Braun** and **Repsold**. Sunrise terminator crosses area from around 11 days to full Moon. Sunset takes place between around 25 days to new Moon.

   Mons Rümker is a shallow-sided plateau with a base diameter of around 70 km that rises like a blister from the marial plains of northern Oceanus Procellarum. It covers an area of 3,800 sq km. When illuminated by a low morning Sun, Mons Rümker takes on a broad, closed crescent appearance, widest in the west. The horns of the crescent in the northeast enclose an area of lower ground. A number of domes, each around 10 km in diameter, emerge on top of the plateau, the highest

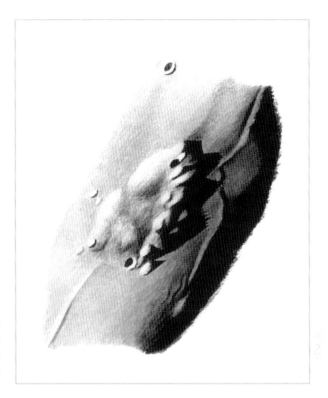

Fig. 7.35. Observational drawing of Mons Rümker, a vast sprawling dome in Oceanus Procellarum. Credit: Grahame Wheatley

of which rise to more than 300 m above the surrounding mare. Mons Rümker may appear large and impressive when it is illuminated by a low Sun, but its slopes rise a gentle average of just 5° from the mare.

Around Mons Rümker, the marial plains are crossed by one or two very inconspicuous wrinkle ridges and are dusted with a few rays, and it is one of the most topographically bland areas on the entire Moon. A wrinkle ridge runs southeast of Mons Rümker to the sharp, bowl-shaped crater **Nielsen** (10 km), just north of the Aristarchus Plateau. Between Nielsen and **Lichtenberg** (20 km), running south of **Naumann** (9.6 km), can be found the wiry wrinkle ridges **Dorsum Scilla** (120 km long) and **Dorsa Whiston** (120 km long). West of Lichtenberg, on the mare border in the libration zone, lies **Ulugh Beigh** (54 km), a flooded, eroded crater whose dark floor can be discerned under a high illumination. A very favorable libration will bring into view the large, far side craters **Röntgen** (126 km) and **Nernst** (118 km), which occupy the eastern floor of the huge basin of **Lorentz** (371 km), due west of Ulugh Beigh. This is one of the biggest far side features brought into view along the libration zone.

Running across the mare immediately northeast from Ulugh Beigh towards the prominent crater **Lavoisier A** (28 km) a chain of sizable flooded craters can be seen. Further north in the libration zone are (from south to north) **Lavoisier** (70 km), **von Braun** (24 km), **Bunsen** (52 km), **Gerard** (90 km) and **Galvani** (80 km), none of which are particularly spectacular. **Repsold** (107 km), on the western shore of Sinus Roris (see Area Six), has one of the most dramatically fractured floors of any lunar crater. **Rimae Repsold** cut extensively across Repsold's floor, the largest of which, a deep linear rille 2 km wide and 120 km long, bisects the crater from the northeast to the southwest into an adjoining crater and the landscape beyond.

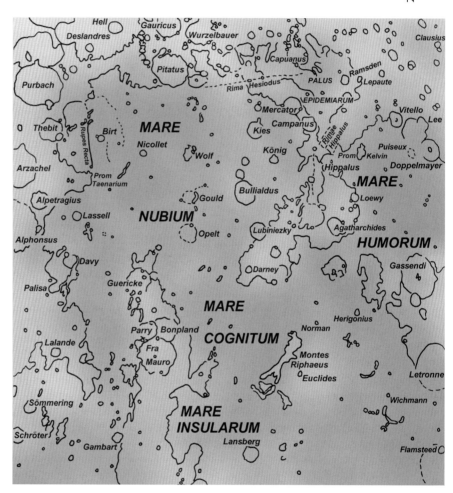

Southwestern Quadrant

Area Nine

Mare Nubium (Sea of Clouds) sprawls across the southeast central part of the Moon. To Mare Nubium's north is **Mare Cognitum** (Known Sea) and the southern reaches of Mare Insularum (see Area Five). Montes Riphaeus, a substantial mountain spine, rises at the junction between Mare Cognitum and Oceanus Procellarum. The prominent joined trio of craters, **Fra Mauro, Bonpland** and **Parry,** are among the flooded features in northern Mare Nubium. Dominating the southwestern mare is the prominent, well-preserved impact crater **Bullialdus,** and near the mare's southeastern border can be found one of the Moon's largest and

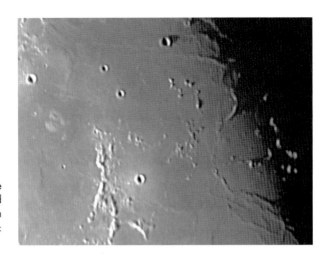

Fig. 7.36. CCD image of Montes Riphaeus and wrinkle features in southern Oceanus Procellarum. Credit: Peter Grego

neatest fault features, **Rupes Recta** (Straight Scarp). **Pitatus** broods over Mare Nubium's southern shoreline. Sunrise over the area is between around 8 and 10 days. Sunset is between around 23 and 25 days.

Lansberg (39 km) is a deep, well-formed crater with broad, internally terraced walls and a cluster of central peaks. It lies along a loosely scattered chain of hills and peaks that runs from Kunowsky (see Area Seven), 200 km to the northwest. A braided wrinkle ridge runs southwest of Lansberg past the flooded crater **Lansberg C** (17 km) and the small, bright-ray crater **Lansberg D** (9.5 km) to converge with several more complex wrinkle ridges in Oceanus Procellarum. A low, irregular domelike swelling can be found immediately east of Lansberg D. To its southeast lie **Montes Riphaeus**, a branching series of mountain ridges 195 km long from north to south, whose southern components mark the northwestern rim of the Mare Cognitum impact basin. Part of the wall of **Euclides P** (60 km), a large, flooded crater on the shore of Mare Insularum, is outlined by the northern ridges of Montes Riphaeus. **Euclides** (12 km), a bright and prominent bowl-shaped crater surrounded by a homogenous collar of bright ejecta material, lies immediately west of Montes Riphaeus. Further west, a scattering of small mountain peaks (unnamed) rises from Oceanus Procellarum; that may be the remnants of a large, buried crater wall.

Mare Cognitum is a dark lava plain, somewhat oval in shape, measuring 330 km from Montes Riphaeus in the northwest to its southeastern shoreline near Guericke. Mare Cognitum, the "Known Sea", takes its name from the fact that the probe, Ranger 7, secured the first detailed close-up photographs of the Moon's surface prior to its (intended) crash-landing on the sea in July 1964. Faint, streaky rays from Copernicus (see Area Five), more than 500 km to its north, can be traced across the mare. Near its western shoreline can be found a teardrop-shaped dome 20 km long. Unique among lunar domes, it is composed of brighter material than the surrounding mare, and it can be seen under a high illumination. A group of hills to its west also have a high albedo. It is possible that the dome and the nearby hills are the remnants of a submerged crater, the dome representing the crater's central uplift. Mare Cognitum's southern border is marked by the small ray crater **Darney** (15 km) and a scattered line of mountains to its east. A narrow wrinkle ridge, 60 km long, joins one of these peaks to **Mons Moro**, a small mountain in the southeastern part of the mare. A number of narrow wrinkle ridges run across other parts of the mare. At the center of Mare Cognitum lies the little bowl-shaped crater **Kuiper** (6.8 km), which shows as a tiny bright dot under a high angle of illumination.

East of Mare Cognitum can be found a collection of large, flooded craters with low, eroded walls. Among them is **Guericke** (58 km), whose rubbly walls are breached to the north and east. Under a low Sun, several domelike swellings can be discerned on its floor. The conjoined trio of **Fra Mauro** (95 km), **Bonpland** (60 km) and **Parry** (48 km) lie on Mare Cognitum's northeastern border, north of Guericke. The mutual wall between the crater trio is higher and more prominent than other parts of the walls, and its shape immediately tells the observer that Fra Mauro is the oldest crater, followed by Bonpland and Parry. The rest of Fra Mauro's wall is low and rather eroded, wrinkled by distinct north–south ridges; this structure can be traced both in the hills to the north of Fra Mauro and across much of its floor, probably part of the Imbrium basin's radial impact sculpting. In February 1971, Apollo 14 landed in the hills just 25 km north of Fra Mauro's wall. One can trace prominent rille running from these hills across the crater wall to the center of Fra Mauro, terminating at the little crater **Fra Mauro E** (4 km). To its south, a small ditch is divided into two prominent rilles. One cuts cleanly across the southeastern wall onto the western floor of Parry, and the other cuts across the southern wall onto Bonpland's northern floor, where it bends west at a sharp angle. Another rille almost bisects Bonpland north–south, and a smaller one runs across the crater's southeastern floor. **Tolansky** (13 km), a short distance south of Parry, is joined to it by a small rille, a continuation of which heads south for 70 km, terminating at the flooded crater **Guericke F** (22 km). Together, the rille system is known as **Rimae Parry**.

A broad, flat, marial plain – an unnamed northern extension of Mare Nubium – occupies an area of some 23,000 sq km east of Parry. Its southern reaches are dented by the deep bowl crater **Kundt** (11 km). To the east is **Davy Y** (64 km), a flooded crater with a distinct rectangular shape. Running across the eastern part of its floor is **Catena Davy**, a crater chain that can be traced 50 km to the bright impact crater of its origin, **Davy G** (12 km). Larger craters along the chain can be resolved in a 150 mm telescope. **Davy** (35 km), a sharp-rimmed, slightly polygonal crater with a bumpy, mottled floor, lies on Davy Y's southwestern wall. Its own southeastern wall is intruded upon by **Davy A** (15 km). A breach in the northern wall of Davy Y leads into the floor of **Palisa** (33 km). The mountains north of Palisa display clear signs of radial Imbrian sculpting. Prominent ridges point north to **Lalande** (24 km), a prominent impact crater, polygonal in outline, with broad internally terraced walls. Lalande lies at the center of a bright ray system, parts of which can be traced to distances of more than 200 km across northern Mare Nubium and southern Mare Insularum.

Mare Nubium is a roughly rectangular sea with an east–west diameter of around 600 km and a total surface area of 254,000 sq km. It is easily visible with the naked eye as a dark patch in the central southern part of the Moon. The ray systems of Copernicus to the north (see Area Five) and Tycho to the south (see Area Ten) overlap in Mare Nubium, but most prominent are Tycho's rays, two of which streak across the western part of the mare. The mountain border of Mare Nubium is incomplete and largely submerged in the north, and where it exists from the west around the south to the east it is indented with craters, many of which are flooded. The western border of Mare Nubium is crossed by the prominent arcuate rille system of **Rimae Hippalus**, which runs concentric to the border of Mare Humorum to the west (see Area Eleven), a branch of which extends northwards across the floor of the disintegrated flooded plain **Agatharchides** (49 km). To its north are several submerged craters, of which **Lubiniezky** (44 km) forms an almost complete

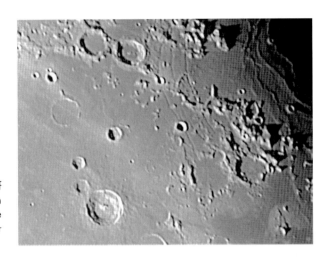

Fig. 7.37. CCD image of western Mare Nubium, with the crater Bullialdus and the Rimae Hippalus. Credit: Peter Grego

narrow ring. Beyond its breached southeastern wall lies **Bullialdus** (61 km), a very prominent crater with broad, intricately terraced inner walls and a large group of central peaks that rises above a smooth floor of impact melt. Bullialdus' rim rises to more than 3,500 m above its floor, and it has a marked indentation along the southeast rim, possibly the result of having impacted so close to the crater **Bullialdus A** (25 km). Bullialdus A is overlain by ejecta and scarred by the impact sculpting that surrounds Bullialdus. At a low angle of illumination, the terrain around Bullialdus appears terrifically detailed, with radial ridges and indications of secondary crater chains that spread in all directions across the mare up to a distance of 60 km from the crater's rim.

König (23 km), a deep-impact crater with a hilly floor, lies halfway between Bullialdus and **Campanus** (48 km) on the southwestern mare border. The thin rim of a submerged crater, **Kies** (44 km), can be seen across the mare southeast of König. Under a low illumination, Kies and König are overlain by one of the most prominent of Tycho's rays. At a low angle of illumination, a number of very interesting features of low relief come into view in the area. Southwest of Kies, a large, round dome, **Kies Pi** (base diameter 12 km) has a small summit crater that is clearly resolvable in a 150 mm telescope. To its west, a low ridge connecting König with Campanus casts a prominent shadow under a low evening Sun, along with another domelike swelling near a scattered group of hills. Campanus and **Mercator** (47 km) are a prominent linked pair on the border between Mare Nubium and the small, irregular plain **Palus Epidemiarum** (see Area Ten). Campanus has more developed internal terracing and a tiny central hill, while Mercator has a smooth, flat floor. Running east from Mercator is **Rupes Mercator** (180 km long), a straight, though somewhat eroded scarp that marks the inner edge of one of the Nubium basin's original mountain rings.

Pitatus (97 km), a large crater on the southern shore of Mare Nubium, has a complex, though rather eroded, wall made up of several low terraces. Its generally flat floor has an offset central elevation, and it is cut across by the complex rille system of **Rimae Pitatus**, which runs around the edge of the floor, in places making its way along parts of the terracing. A 100 mm telescope can resolve many of these rilles. Adjoining Pitatus in the west is the flooded plain **Hesiodus** (43 km). To its south, **Hesiodus A** (15 km), a crater with a beautiful complete inner ring – one of the nicest examples on the Moon, though somewhat on the small side, best viewed at high magnification through a 100 mm telescope. Further west, one of the widest

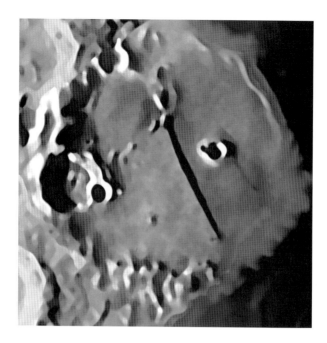

Fig. 7.38. Observational drawing of Rupes Recta in Mare Nubium. Credit: Peter Grego

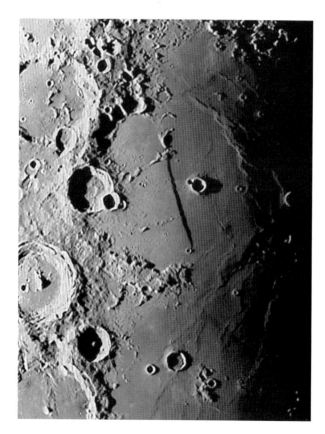

Fig. 7.39. Rupes Recta, a large fault in Mare Nubium. Credit: Mike Brown

and longest linear rilles on the Moon, **Rima Hesiodus** (300 km long) runs westwards into Palus Epidemiarum (see Area Ten).

Running down the center of Mare Nubium, east of Bullialdus, a chain of six flooded craters can be discerned, outlined in places by solid remnants of crater rims intertwined with wrinkle ridges. The chain proceeds south from an unnamed bay north of **Opelt** (49 km), through **Gould** (34 km), to the blocky cluster of peaks outlining **Wolf** (25 km) in the south – a distance of 285 km. These features are aligned radially to the Imbrian basin, and it is possible that they represent a submerged crater chain that was produced during the Imbrian impact. Under a high angle of illumination, close examination will reveal the darker flooded floors and some of the brighter walls along the crater chain, along with the bright peaks surrounding Wolf.

Several wrinkle ridges in eastern Mare Nubium converge in the vicinity of the crater **Nicollet** (15 km). The ridges in the east appear to mark the position of the submerged western wall of a large, unnamed crater, around 200 km in diameter, whose eastern wall makes a deep semicircular bay on the eastern coastline of Mare Nubium. **Promontorium Taenarium**, a broad, squared-off headland, projects westwards along the northern part of the bay. To its south, and almost completely linking the two opposite sides of the bay, is the magnificent scarp face of **Rupes Recta** (110 km long). Sometimes known as the "Straight Wall" – though it is not perfectly straight, nor can it be considered a wall – Rupes Recta can be observed

Fig. 7.40. Rupes Recta, a large fault in Mare Nubium.
Credit: Mike Brown

through telescopes as small as 60mm. Illuminated by a rising Sun, the fault casts a prominent broad, dark shadow onto the mare. Under an evening illumination, the scarp face can be seen as a prominent thin bright line, but no trace of it can be discerned under a high angle of illumination. Rupes Recta is the finest and most clear-cut example of a normal fault on the Moon. After this part of Mare Nubium was flooded with lava, tension in the crust caused it to crack, and the western side of the fault has dropped down up to 300 meters. Through a telescope the scarp may appear as a steep slope, but it is not as precipitous as appearances might suggest: it has a gradient of about 7°, so gentle that it could be ascended without much difficulty. A small crater, **Arzachel D** (5 km), lies at the fault's northern end. In the south, the fault cuts through the southern components of a group of small peaks known (unofficially) as the "Stag's Horn Mountains". These mountains mark part of the disintegrated western rim of a flooded crater, **Thebit P** (65 km), whose western floor displays a distinct cluster of small, rounded dark patches.

Some 20 km west of Rupes Recta lies the prominent, deep bowl-shaped crater **Birt** (17 km), 3,400 m deep, whose eastern rim is overlain by **Birt A** (6 km). An unusual pattern of light gray rays from Birt can be traced across the mare, but in places it is difficult to differentiate some of these from the rays of Tycho. **Rima Birt**, a prominent sinuous rille, commences from the summit crater of a low dome, the elongated crater **Birt E** (9 km long), and runs southwards for 50 km across the mare, terminating at the small crater **Birt F** (2.5 km) on Birt's western flanks. Rima Birt can be resolved through a 100 mm telescope under ideal conditions, though it requires a 150mm telescope to discern clearly. East of Rupes Recta, on the shore of the bay, the prominent crater **Thebit** (57 km) displays a clear-cut rim and well-defined internal terracing, though its hilly floor lacks a central elevation. **Thebit A** (20 km) is superimposed upon Thebit's western rim, and the western rim of Thebit A itself is overlain by a smaller crater, **Thebit L** (10 km). Close scrutiny will reveal a small central peak in the latter.

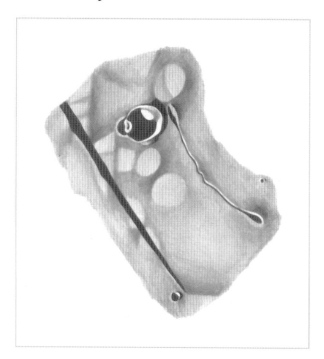

Fig. 7.41. Observational drawing of Rupes Recta and the narrow Rima Birt. Credit: Peter Grego

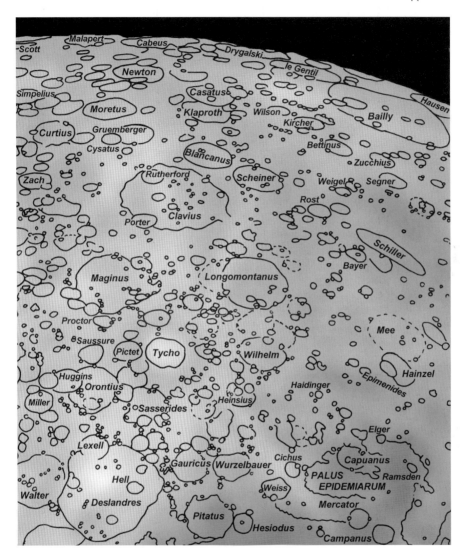

Area Ten

This crater-crowded area of the southern lunar uplands south of Mare Nubium to the limb is dominated by several very large craters: **Deslandres** in the northeast, **Clavius** in the south and **Bailly** near the southwestern limb (the south polar region and its environs are described in Area Thirteen). Between Clavius and Deslandres lies the magnificent crater **Tycho**, at the center of the Moon's brightest and most extensive ray system. **Palus Epidemiarum** (Marsh of Epidemics) and **Lacus**

Fig. 7.42. CCD image of the large eroded crater Deslandres. Credit: Peter Grego

Timoris (Lake of Fear), both in the northwest, represent the only patches of mare in the area. In the far west is the large and ancient **Schiller-Zucchius** impact basin. Sunrise terminator sweeps across the area between around 7 and 11 days. Sunset takes place between around 22 and 26 days.

Lying just southeast of Mare Nubium (see Area Nine), **Deslandres** (235 km) is one of the Moon's largest craters. It is an ancient feature whose walls are low and considerably eroded, but its wall can clearly be traced when illuminated by a low Sun. The outline of Deslandres is, however, quite unrecognizable under a high illumination. The eastern side of its wall is overlain by the large crater **Walter** (see Area Fourteen) and **Walter W** (31 km). A neat linked chain of six craters (from 6 to 4 km in diameter) crosses the northeastern part of Deslandres' floor. The sharp-rimmed crater **Hell** (33 km) lies on Deslandres' hilly western floor – despite its name, this particular Hell is only 2,200 m deep, with a dark floor composed of a mass of ridges and an offset, somewhat pyramidal, central peak. Hell lies along a small arcuate ridge that may represent part of the rim of a buried inner ring of Deslandres. **Lexell** (63 km) forms a prominent bay in Deslandres' southeastern wall, and to its north lies a little crater (unnamed), 3 km wide, that lies at the center of a prominent splash of bright ray material that extends 30 km across Deslandres' floor.

Two large craters, **Gauricus** (79 km) and **Wurzelbauer** (88 km), lie west of Deslandres, to the south of Pitatus (see Area Nine). Gauricus has broad inner walls, peppered with craters of various sizes, and it has a rather smooth, slightly convex floor, with no trace of a central uplift. Wurzelbauer, to its immediate west, is a considerably more eroded crater, with an arc of low mountains that runs across its floor. A bright patch to its west surrounds the small, sharp-rimmed crater **Cichus B** (13 km), made all the more noticeable by being located near the path of a bright ray from Tycho, some 320 km to its southeast. The ray crosses the eastern flanks of **Cichus** (41 km), which lies on the eastern shore of **Palus Epidemiarum**. Measuring 300 km from east to west, with a surface area of some 27,000 sq km, Palus Epidemiarum is an irregular-shaped patch of mare that lies to the immediate southwest of Mare Nubium. The western part of **Rima Hesiodus** (see also Area Nine), one of the Moon's finest linear rilles, cuts across Palus Epidemiarum's northeastern wall and across the northern part of its floor – an easy object to discern through a 100mm telescope. South of the rille, **Capuanus** (60 km) is a flooded crater whose wall is broader and higher in the west; its narrow eastern rim has only just avoided being submerged completely by mare material. Several low, elongated domelike swellings can be observed on the crater's floor. South of Capuanus, **Capuanus P** (84 km long, 35 km wide) is a valley flooded by mare material. In western Palus Epidemiarum, the sharp-rimmed flooded crater **Ramsden** (25 km) is associated with **Rimae Ramsden**, a superb collection of interconnected linear rilles that lie primarily to the east of the crater. Rills run from the northern, southern and eastern flanks of Ramsden, but the rim and floor of the crater remain untouched; Ramsden was clearly formed after the crust had faulted. The main north–south component of the system crosses to the small crater **Marth** (9 km), a beautiful and rare example of a crater with a complete inner ring (like Hesiodus A – see Area Nine). The rille curves around to the mountain border in the north, and another can be seen on the plain just north of Marth. Combined, the rilles of Rimae Ramsden visible in Palus Epidemiarum have a total length of more than 300 km, and they are easy to resolve through a 150mm telescope.

A rille branching from the Rimae Ramsden system cuts through the mountains south of Palus Epidemiarum and west of **Elger** (21 km), southwards over a distance of 125 km to the ramparts of **Hainzel A** (53 km), a prominent crater with a broad, terraced inner wall and a central mountain. A breach in the southeastern wall of Hainzel A leads to the disintegrated, somewhat irregular crater **Hainzel C** (44 km) and the deep elongated trough of **Hainzel** (53 km long) to the south. This complex, interlocked trio makes up a feature some 92 km long from north to south, which itself overlies the northern wall of **Mee** (132 km), an ancient and substantially eroded crater with an irregular rim and a rough, blocky floor. **Lacus Timoris**, a smooth, dark lava plain with an irregular border, lies to the northeast of Hainzel. In the vicinity there are numerous flooded craters, one of the most unusual being the elongated crater **Schiller** (179 km long, 71 km wide). Schiller's broader southern floor is smooth, but the northern half of its floor has a prominent central mountain ridge. A very large unnamed impact basin occupies the area between Schiller and **Zucchius** (64 km); the **Schiller-Zucchius** basin, as it is known, has an outer diameter of approximately 380 km and a flooded inner ring around 210 km across. **Schickard** and the linked trio of **Phocylides**, **Nasmyth** and **Wargentin** (see Area Twelve) lie just beyond the western margin of the basin's outer ring. Best preserved is the far southeastern part of the basin's outer wall, an arc of ridges that run from the smooth-floored crater **Rost** (49 km) to the northern flanks of Zucchius, a prominent deep crater with a broad, terraced inner western wall and a small group of central peaks. **Segner** (67 km), a low-walled crater with a floor crossed by ridges emanating from Zucchius to its south, straddles the inner ring of the Schiller-Zucchius basin. West of Segner, the inner ring can be traced along a narrow, disjointed ridge for around 90 km. East of Segner, the inner ring is composed of a broad series of ridges overlain with craters, including the flooded **Weigel** (36 km) and **Weigel B** (34 km), curving around to the southern flanks of Schiller. Under a high angle of illumination, the inner part of the Schiller-Zucchius basin can be discerned through binoculars as a dark gray patch (with a darkness about halfway between a mare and upland area) near the southwestern limb, surrounded by a less prominent and less well-defined dusky gray zone.

Beyond the Schiller-Zucchius basin, in the libration zone near the southwestern limb, lies **Bailly** (305 km), a sizable multiringed impact basin whose low, eroded outer wall encloses a rough and cratered floor that displays traces of an inner ring (about 150 km in diameter) and linear ridges aligned with the Orientale impact

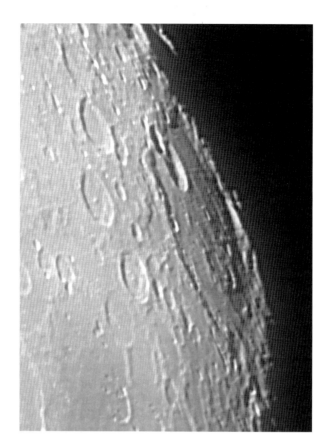

Fig. 7.43. CCD image of
the vast crater Bailly. Credit:
Peter Grego

basin, around 1,000 km to its north (see Area Eleven). On the southeastern floor, **Bailly B** (58 km) is overlain by **Bailly A** (38 km). A little further south along the limb, the highly eroded crater **le Gentil** (113 km) can also be discerned during a favorable libration. Further west of Bailly, lying along the 90°W line, a good libration will also bring into view the prominent deep crater **Hausen** (167 km), with its wide, intricately terraced inner walls and flat knobbly floor overseen by a large group of central mountains that is somewhat offset to the east of the floor.

Tycho (85 km), one of the Moon's best preserved major impact craters, lies at the center of the Moon's largest and most prominent system of rays. Blasted out of the Moon's southern uplands, it is an impressive enough crater as it is, with a sharp rim and wide, intricately terraced inner walls that surround a floor (4,800 m deep) filled with a pool of impact melt and dominated by a large pair of central mountains. Outside Tycho's rim, a predominantly concentric set of ridges can be traced to a distance of around 35 km, but some radial structure can be discerned, including several crater chains to its northwest. Under a high Sun, Tycho's rim and central peaks show up as bright features surrounded entirely by a dark collar of impact melt. Further afield, the bright rays commence their long paths in all directions across the Moon's surface. Binoculars will reveal three particularly bright rays – one travels southwest towards the limb, and the others form a close parallel pair that extend northwest over the highlands and across the western reaches of Mare Nubium. Many somewhat less bright rays splash out to the north and east of Tycho,

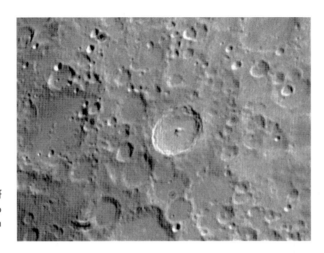

Fig. 7.44. CCD image of the bright ray crater Tycho in the Moon's southern uplands. Credit: Peter Grego

covering much of the southeastern quadrant of the Moon to distances of more than 1,700 km. Close scrutiny of these rays under a high illumination will reveal that many of them are not continuous lines, but are instead composed of a multitude of short streaks, typically around 20 km long. Some observers receive the distinct impression that a prominent ray that bisects Mare Serenitatis is one of Tycho's. This is an illusion, since the ray in question actually emanates from Menelaus on the mare's southern shore (see Area One), although some of Tycho's rays can be traced in the area south of Mare Serenitatis, and some ray material does appear to traverse the dark plains of Mare Nectaris (see Area Fifteen), 1,300 km from Tycho.

Tycho is surrounded by a number of prominent craters. To the northeast lies **Orontius** (122 km), a plain with a considerably eroded wall whose eastern floor rises up to meet the rim of **Huggins** (65 km), a prominent crater with a large central mountain. Huggins is crescent-shaped bcause its eastern wall is overlain by **Nasireddin** (52 km). Adjoining Nasireddin's northern wall is **Miller** (75 km). Both Nasireddin and Miller have well-developed terracing and central peaks surrounded by smooth impact melt. A prominent, broad bulging ridge pushes north from Nasireddin across Miller's floor, almost touching its central peaks – an unusual feature that is too substantial to represent a landslide. South of Orontius, flat-floored **Saussure** (54 km) overlies a larger, unnamed crater whose sharp rim can be seen to the east. **Pictet** (62 km), a crater with an eroded rim, lies immediately east of Tycho and is somewhat overshadowed by it. Similarly eroded **Pictet E** (55 km), to its north, joins the wall of **Sasserides** (90 km), an ancient disintegrated feature, much of whose walls are overlain with younger craters. Under a low Sun, the floors of Sasserides, Pictet and Pictet E display considerable grooving radial to Tycho.

An arc of ten craters runs from **Tycho A** (25 km), 50 km north of Tycho, around to **Tycho X** (11 km) in the west – the craters are roughly parallel with Tycho's rim. Further to the northwest, a line of sizable craters runs south from **Heinsius** (20 km) to the northeastern ramparts of **Wilhelm** (107 km), a large plain with a broad inner western wall, its eastern floor dotted with a number of smaller craters. The eroded, irregular shaped crater **Montanari** (77 km) to its south links Wilhelm with the prominent crater **Longomontanus** (145 km), one of the largest craters

Fig. 7.45. CCD image of the large crater Clavius in the Moon's southern uplands. Credit: Peter Grego

in the southern uplands. Longomontanus's broad walls are dotted with small craters and have a spectacular sculpted appearance. Its floor is a broad plain of impact melt, from which rises a small cluster of near-central peaks. Longomontanus overlies the western half of a smaller crater, **Longomontanus Z** (77 km). **Maginus** (163 km), southeast of Tycho, is rather similar in appearance to Longomontanus and is equally impressive.

Clavius (225 km), a giant crater that is absolutely stunning to behold through any-sized telescope, is the chief topographic attraction of the Moon's southern uplands. Dented in places by large craters, Clavius's scalloped rim bounds a substantial, blocky inner wall. **Porter** (52 km), a sharp-rimmed crater with a central mountain ridge, is superimposed on Clavius's northeastern rim. **Rutherford** (50 km) lies just within Clavius's southeastern rim, and has an interestingly arranged group of large central mountains. Traces of the external radial impact sculpting of both Porter and Rutherford can be observed on Clavius's floor. An arc of individual craters of diminishing size proceeds from Rutherford westwards across Clavius's floor: **Clavius D** (28 km), **Clavius C** (22 km), **Clavius N** (13 km), **Clavius J** (11 km) and **Clavius JA** (8 km). A group of peaks representing

Fig. 7.46. CCD image of Clavius in the late afternoon. Credit: Peter Grego

Clavius's central uplift lies immediately southwest of Clavius C. Lying just 320 km due south of Tycho, Clavius has not escaped being dusted with Tycho's ejecta, and a couple of light rays can be traced running southwards across its floor.

Southwest of Clavius, **Blancanus** (105 km) and **Scheiner** (110 km) make an interesting pair to compare. Blancanus is clearly the younger feature, with a clearly defined circular rim and a regular series of inner terraces. Its floor is smooth, except in the south, where there is a small group of hills and a cluster of small craters. Scheiner, to its immediate northwest, is a considerably more ancient feature, its wall and floor having been battered and eroded by multiple impacts. South of Blancanus, the conjoined duo of **Klaproth** (119 km) and **Casatus** (111 km) are an easily recognizable landmark, one of the pointers to the Moon's south polar region. Klaproth's smooth, gray floor is surrounded by a low wall, which is shared by Casatus to its south. The remainder of Casatus's wall is broader and more sharply defined than that of Klaproth, and its floor has a prominent bowl-shaped crater, **Casatus C** (15 km), just north of center. The large crater **Drygalski** (163 km) lies in the libration zone along the limb south of Casatus. Drygalski straddles the 90°W line of longitude, just 170 km from the south pole, and can be glimpsed at a favorable libration and illumination.

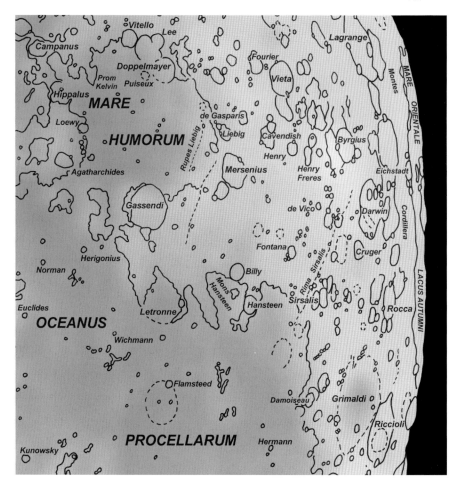

Area Eleven

Much of this area is taken up by the southern plains of Oceanus Procellarum, its jagged, irregular shoreline indented here and there with flooded craters; the largest of these, **Letronne**, forms a prominent bay on the southern coast. Further south is the impressive crater **Gassendi**, which lies on the northern shore of the dark, circular sea **Mare Humorum**. West of Oceanus Procellarum lies **Grimaldi**, a large, dark, circular plain. To its southwest, in the libration zone along the western lunar limb, lies **Mare Orientale** (Eastern Sea), which occupies the central ring of the Orientale impact basin. Parts of the basin's outer rings are marked by **Montes Rook** and **Montes Cordillera**, between which lie the narrow lava patches of **Lacus Veris** (Lake of Spring) and **Lacus Autumni** (Lake of Autumn). **Byrgius A**, between Mare

Humorum and Mare Orientale, adds a prominent splash of rays to the cratered highlands. Sunrise over the area is between around 10 days to full Moon. Sunset is from around 25 days to new Moon.

Wrinkle ridges, small, bright, bowl-shaped craters with ejecta collars, scattered mountains and the remnants of flooded craters litter the southern reaches of Oceanus Procellarum. **Flamsteed** (21 km) lies on the southern wall of a large, almost completely submerged crater, **Flamsteed P** (107 km), traces of whose rim can be discerned along some narrow arcs of hills and low ridges. East of Flamsteed lies the bright, bowl-shaped crater **Wichmann** (11 km), which sits on the south-eastern wall of the flooded **Wichmann R** (52 km), whose northern rim protrudes in the shape of a brandy glass. West of Flamsteed is a conjoined group of five flooded craters of various sizes, the largest of which is **Flamsteed G** (55 km). Although these are prominent when illuminated by a low morning or evening Sun, the rims of all these flooded features in the Flamsteed area can easily be seen as light gray outlines on the darker mare. A number of wrinkle ridges proceed south of Flamsteed, notably **Dorsa Rubey** (100 km), a component of which reaches across the mare to the small group of hills that arcs the remnants of the central elevation of **Letronne** (119 km). Letronne forms a prominent bay in the southern border of Oceanus Procellarum, whose western wall juts out to create a narrow headland along which lies **Winthrop** (18 km), a flooded crater. Two unusual elongated craters, **Letronne X** (21 km long) and **Letronne W** (15 km long), lie within Letronne's eastern wall. A small, bright impact crater with an ejecta collar, **Letronne B** (7 km) is located near the southeastern edge of Letronne's floor. East of Letronne, a broad and complex system of wrinkle ridges, **Dorsa Ewing**, heads 200 km east near the mare border across to **Herigonius** (15 km), from which more ridges head further east. Two branches of **Rima Herigonius**, a narrow, sinuous rille system, originate on two separate ridges of Dorsa Ewing to join up in the south, heading for the mountains southwest of Herigonius – a challenge to resolve in a 200mm telescope.

West of Letronne, a short distance along the mare border, lie an interesting, close crater duo, **Billy** (46 km) and **Hansteen** (45 km). Billy is a sharp-rimmed circular crater with a low wall that borders a very dark plain that appears perfectly smooth through a telescope smaller than 100mm. Larger instruments may reveal a little irregularity on Billy's floor when it is lit by a low Sun, notably in the south, where there are a couple of very small craters surrounded by gray ejecta collars. Hansteen, northwest of Billy, is slightly polygonal in outline, with a sharp rim that surrounds a complex floor with a distinctly concentric series of hills and ridges. **Rima Hansteen** (25 km), a narrow, sinuous rille, runs along the western flanks of Hansteen. Under a high Sun, a dark, irregular lava patch can be seen on Hansteen's northern floor. A sizable irregular bay to the east of Hansteen is dominated by **Mons Hansteen**, a large mountain massif with a triangular base measuring 30 km across; it shows up as a brilliant patch under a high angle of illumination, contrasting nicely with the dark patch of Billy to its south. Another dark, irregular plain lies south of Billy, and the dark-floored crater **Zupus** (38 km), with its somewhat eroded walls, can be found on the southern border of this unnamed plain. A number of narrow linear rilles, **Rimae Zupus**, lie to its northwest.

Grimaldi (222 km), a broad, dark lava plain between the border of Oceanus Procellarum and the western limb of the Moon, is one of the most prominent features in the area. Binocular users will easily locate the feature any time that it is illuminated, regardless of the effects of libration, and exceptionally keen-sighted

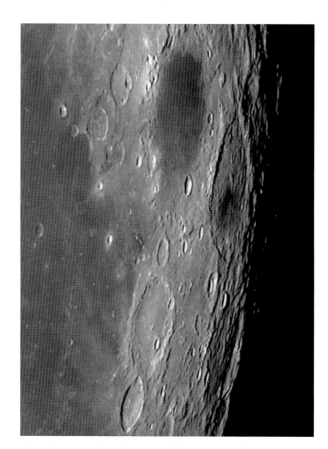

Fig. 7.47. The large dark floored crater Grimaldi near the Moon's western limb. Credit: Mike Brown

people claim the ability to discern Grimaldi without optical aid. Grimaldi's dark floor is a little lighter in tone in the north, where it is dusted with ray material from Olbers A, more than 400 km to the north (see Area Seven). Illuminated by a low morning Sun, a large, low dome with a base diameter of around 20 km can be seen on the northern part of the floor, along with several low ridges that run southwards across the floor. The hills surrounding Grimaldi are complex, lined with numerous faults and rilles. **Rimae Grimaldi**, a system of linear rilles to the east and southeast of Grimaldi, can be traced from the west of the complex multiple crater **Damoiseau** (37 km) southwards across the uplands into **Sirsalis Z** (80 km), a total distance of around 250 km. When Grimaldi has just emerged into the morning sunlight, it is possible to trace the larger impact basin (430 km in diameter) of which Grimaldi is central. An arc of low ridges from Damoiseau to **Rocca** (90 km) in the south marks the most visible part of the outer wall of the Grimaldi impact basin.

Immediately northwest of Grimaldi, the large crater **Riccioli** (146 km) has a well-defined low wall and a rough, hilly floor that is cut through by the numerous linear rilles of **Rimae Riccioli**. The northern part of Riccioli's floor is relatively smooth and dark, the lava having submerged traces of the Rimae Riccioli beneath its surface. Additionally, the whole of Riccioli and the surrounding area is overlain with a texture radial to the Orientale impact basin to its south. Near the limb southwest of Riccioli, in the libration zone, the prominent crater **Schlüter** (89 km) has a strongly terraced internal wall and a large central peak.

Fig. 7.48. Mare Orientale and surrounding lava lakes imaged at a favorable libration. Credit: Peter Grego

Schlüter lies in the northeastern part of **Montes Cordillera**, a vast circular mountain range with peaks rising to heights of more than 5,000 m, marking the 930-km-diameter outer ring of the vast Orientale multiringed impact basin – the Moon's youngest major asteroidal impact scar at around 3.2 billion years old. The topographic effects caused by the Orientale impact don't end at the Montes Cordillera; radial structure, in the form of ridges, grooves and chain craters, can be clearly traced for hundreds of kilometers further across the Moon's surface. This radial structure is most noticeable across parts of the far side where there has been little flooding by subsequent lava flows, notably Vallis Bouvard, Vallis Baade and Vallis Inghirami (see Area Twelve), near the southwestern limb. There are some indications of further Orientale ring structures measuring some 1,300 and 1,900 km in diameter. **Eichstadt** (49 km), a prominent crater on the eastern edge of Montes Cordillera, lies at 78°W, on the limit of the libration zone. At an extreme libration west, Eichstadt and the Cordilleras in its vicinity remain visible in profile along the limb, while the rest of the basin lies beyond the limb.

A broad, hummocky plain lies between Montes Cordillera and **Montes Rook,** a 620-km-diameter inner mountain ring of the Orientale basin. In the northeast, the plain is stained by **Lacus Autumni**, a small collection of irregularly shaped dark lava patches that cover a total surface area of around 3,000 sq km. Inside Montes Rook, the long and narrow dark lava plains of **Lacus Veris** cover an area of around 12,000 sq km. A jumbled hilly plateau around 100 km wide lies between Lacus Veris and **Mare Orientale** itself. At 300 km wide, this circular mare is the dark bull's-eye of the Orientale multiringed impact basin. Although the eastern border of Mare Orientale lies on the 90°W line of longitude, all of Mare Orientale and some of the hills to its west lie within the libration zone, so at times it can be seen in its entirety as an extremely foreshortened dark line near the limb. Two large craters lie on the margins of Mare Orientale – **Maunder** (55 km) to the north and **Kopf** (42 km) to the east. In the northern central part of the mare is **Hohmann** (16 km), and to its west lies the small crater **Il'in** (13 km).

Cruger (46 km), a circular crater with a smooth, dark floor, occupies the center of a larger, less distinct (unnamed) crater around 120 km in diameter whose southwestern wall joins with the crater **Darwin** (130 km). A tiny crater lies at the center of Cruger, visible through a 150mm telescope. Nearby, **Lacus Aestatis** (Lake of

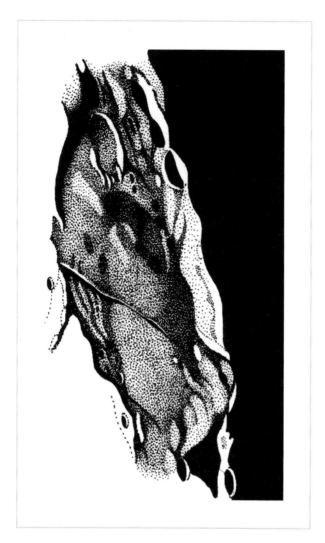

Fig. 7.49. Stippled observational drawing of Darwin, a large and very ancient crater near the Moon's western limb. Credit: Nigel Longshaw

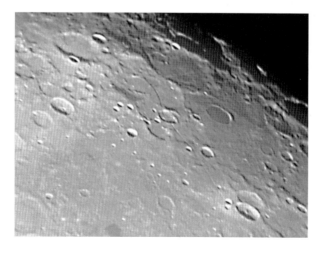

Fig. 7.50. CCD image of the Rima Sirsalis area. Credit: Peter Grego

Summer) is composed of two unconnected dark lava flows that occupy part of the floor of **Rocca A** (64 km) and a small area north of Cruger. A number of smaller dark patches can be discerned between Lacus Aestatis and **Sirsalis** (42 km), a prominent crater with a terraced inner wall and a sizable central peak. Sirsalis overlaps the slightly larger and more eroded **Sirsalis A** (45 km). To their east, beginning at the very southwestern shoreline of Oceanus Procellarum, the prominent linear rille **Rima Sirsalis** (400 km long) cuts southwards across the highlands and older craters, across the floor of a large unnamed crater (119 km) east of the large, eroded crater Darwin, cutting at right angles through the more ancient linear rilles of **Rimae Darwin**. Several younger craters – the small, bright ray crater **Sirsalis F** (12 km), for example – are superimposed on Rima Sirsalis. Several smaller rilles in the Sirsalis system can be found in the hills bordering Oceanus Procellarum, but the main rille itself is easily visible through a 100 mm telescope. South of Darwin, overlapped by its southern wall, is the highly eroded crater **Lamarck** (115 km). To its east is **Byrgius** (87 km), whose eastern rim is overlain by **Byrgius A** (17 km), a small, bright crater at the center of a prominent splash of rays, easily visible through binoculars, which reach distances of more than 300 km.

Twin craters **Henry** (41 km) and **Henry Frères** (42 km) are coated with bright ejecta from Byrgius A, which lies a short distance to their west. To their east lies **Cavendish** (56 km), a sharp-rimmed crater whose floor displays the remnants of two sizable (15 km-diameter) submerged craters. Cavendish's southwestern rim is overlain by **Cavendish E** (26 km), and its eastern flanks are intruded upon by components of **Rimae de Gasparis**, a complex system of linear rilles that both surrounds **de Gasparis** (30 km) and cuts across its floor. The system resembles Ramsden and Rimae Ramsden (see Area Nine), 540 km away on the other side of Mare Humorum, except that the floor of de Gasparis is crossed by rilles, while Ramsden's floor is rille-free. Around half a dozen interlocking linear rilles running across an area of around 130 km in diameter between Cavendish, **de Gasparis A** (32 km) and **Liebig** (37 km), make up the complex local network of Rimae de Gasparis. But the rilles are part of a bigger picture of faulting consequent upon the enormous crustal adjustments in and around the Humorum impact basin, traceable in numerous linear and arcuate rilles that run extensively through the hills, mountains and craters west and east of Mare Humorum.

Mare Humorum, a prominent circular sea 410 km in diameter, occupies an area of around 120,000 sq km. It has a well-defined border except in the northeast, where its wall is partly submerged by lava flows linking Mare Humorum with the southern plains of Oceanus Procellarum. The original Humorum multiring impact basin was a much more extensive entity, but most traces of its outer rings have been obliterated by subsequent impacts and flooding by lava. For example, parts of an 800-km-diameter Humorum mountain ring can be traced in the area southeast of Billy; they are surrounded by darker mare-filled plains in the vicinity that have been largely obscured by subsequent layers of ejecta. Southwest of Mare Humorum, **Palmieri** (41 km), a flooded tadpole-shaped crater which is crossed by the linear rilles of **Rimae Palmieri**, lies at the southern end of a dark lava plain (120 km long) that represents a flooded part of the broader Humorum basin.

Mare Humorum's interior is dotted with small, bright craters, and it is more uniform, tone than most other lunar maria. It is generally of lighter in the south, with a well-defined darker tract of mare material near the southern shoreline. Ray material can be traced around **Doppelmayer K** (6 km) and a few small unnamed craters, along with several indistinct linear streaks in the western half of the mare.

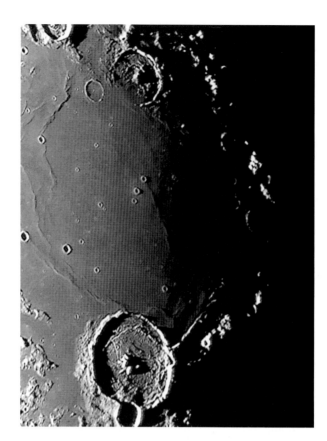

Fig. 7.51. Mare Humorum.
Credit: Mike Brown

A well-defined concentric system of wrinkle ridges runs entirely around the eastern part of Mare Humorum, perhaps marking the location of the complex eastern rim of a submerged inner mountain ring more than 200 km in diameter. However, it is unusual that wrinkle ridges are absent in the west.

Gassendi (110 km), a magnificent crater, dominates the landscape north of Mare Humorum. Much of the crater overlies the mountainous plateau to the

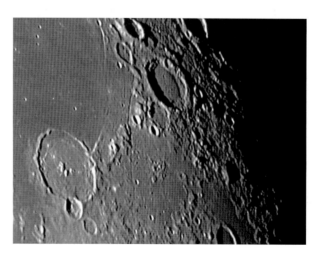

Fig. 7.52. Gassendi lies on
the shore of Mare Humorum.
Credit: Brian Jeffrey

north, but its southern wall projects into northern Mare Humorum, where it narrows as its southern rim becomes almost submerged. Gassendi's floor is a complex collection of hills, mountains, ridges and linear rilles. A group of three large central mountains is surrounded by **Rimae Gassendi**, which cut across much of the crater's floor, mostly to the east of the central peaks. Resolvable thorough a 100 mm telescope, the most prominent rille proceeds east of the largest central peak and curves towards the crater's eastern wall before heading south, just inside a low, concentric ridge that marks a submerged inner wall of Gassendi's. The other rilles require a 150mm telescope to resolve adequately. Added together, the largest components of Rimae Gassendi would stretch for more than 300 km. **Gassendi A** (33 km), a sharp-rimmed crater with a central spine of hills, is superimposed upon Gassendi's northern rim. A ridge to its south extends a short distance across Gassendi's floor, causing a prominent triangular shadow to be cast at a morning illumination. A sharply defined east-facing scarp proceeds for a distance of 60 km from Gassendi's southwestern wall along the mare shoreline; there it is gradually submerged. Its course is taken up further south by a component of Rimae Gassendi, which itself transforms into another prominent east-facing scarp, **Rupes Liebig** (180 km long).

Mersenius (84 km), on the western shore of Mare Humorum, is a prominent crater with a broad, terraced wall surrounding a smooth and markedly convex floor. An illusion causes the convex appearance of Mersenius's floor to be exaggerated, since the floor's albedo is considerably lower in the west, along the base of the

Fig. 7.53. Showing the extensive faults on the floor of Gassendi. Credit: Mike Brown

inner western wall, while the eastern part of the floor is dusted with some lighter ejecta emanating from **Mersenius C** (12 km) to the northeast. Under a high morning Sun, the crater's floor can therefore appear markedly convex through an albedo effect alone. A 150 mm telescope will resolve a line of small craters that runs south–north along the middle of Mersenius's floor. East of Mersenius, several linear riles of **Rimae Mersenius** score the shoreline of Mare Humorum, the longest of which runs for 250 km, beginning north of the flooded crater **Mersenius D** (30 km) and ending in the mountains west of Gassendi. **Vieta** (87 km), 230 km south of Mersenius, also has broad-terraced inner walls that surround a markedly convex floor, and it too displays a line of small craters. **Fourier** (52 km) lies to its east, and a large teardrop-shaped dome topped by a small summit crater lies immediately south of Fourier.

Doppelmayer (64 km), a large, partly submerged, crater lies on the southern coast of Mare Humorum. Much of its northern rim has been submerged by mare lava flows, which have spread onto the northern part of the crater's floor. Doppelmayer's central peak is large and prominent, linked to a ridge that runs around the western part of the crater floor. Under a low angle of illumination, Doppelmayer can appear to be a complete formation, with a prominent inner ring, its submerged northern rim traceable by the shadows cast by a low mare ridge. West of Doppelmayer, the narrow rilles of **Rimae Doppelmayer** (130 km long) can only be resolved with a 150 mm telescope under ideal conditions. East of Doppelmayer, **Puiseux** (25 km) is almost completely submerged beneath the mare, its remaining rim marked by a narrow ring of hills surrounding a smooth plain from which pokes a tiny central hill. East of Puiseux, the wrinkle ridges in southern Mare Humorum converge and firm up into a narrow mountain ridge that marks part of the northern wall of the flooded bay crater **Lee M** (51 km), which is linked to another flooded bay crater **Lee** (41 km) on the shoreline southeast of Doppelmayer. To their east, **Vitello** (42 km) is a prominent fully formed crater with a sizable central mountain surrounded almost entirely by a small, sinuous rille.

Mare Humorum's southeastern border is made up of a prominent range of mountains, the western edge of which, **Rupes Kelvin**, makes a straight 190-km-long scarp face, interrupted at intervals by small, narrow headlands that jut out into the mare, rather reminiscent of coastal groins. **Promontorium Kelvin**, a huge moun-

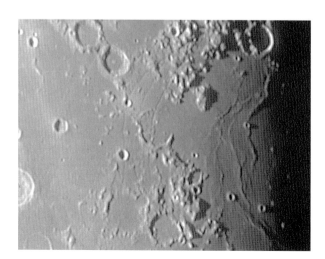

Fig. 7.54. CCD image of Hippalus and Rimae Hippalus, arcuate rilles near Mare Humorum. Credit: Peter Grego

tain mass with faceted sides and a triangular base some 35 km wide, lies north of Rupes Kelvin. **Rimae Hippalus** (see also Area Nine), a prominent system of arcuate rilles, run parallel to the eastern border of Mare Humorum. More than 300 km long, the most prominent of these rilles cuts through the mountains east of Rupes Kelvin and proceeds north across the floor of the flooded bay crater **Hippalus** (58 km), terminating south of Agatharchides (see Area Nine). At least four major arcuate rilles of Rimae Hippalus can be discerned through a 100mm telescope, and traces of smaller associated rilles can be resolved in many nearby parts of the mare border. **Loewy** (24 km) is one of a number of small, flooded craters that mark the northeastern border of Mare Humorum. Between Loewy and Gassendi, a few scattered mountain peaks puncture the smooth plains that link across to Oceanus Procellarum. Several low ridges and large, irregular domelike swellings can also be seen at a low angle of illumination; these include a very large domelike plateau north of Agatharchides, 56 km in diameter, punctured by sharp mountain peaks and overlain by a branching ridge.

Area Twelve

Schickard, a very large crater with a flat, multitoned floor, dominates this part of the southwestern near-limb area. To its north lies the large, irregular dark lava plain of **Lacus Excellentiae** (Lake of Excellence). Of considerable observational interest is the connected trio of craters, **Phocylides**, **Nasmyth** and **Wargentin**, south of Schickard. Wargentin is a crater that appears to have been filled almost to the rim with lava, and it resembles a rather flat circular plateau. Large, near-limb craters include the somewhat disintegrated **Lagrange** and **Piazzi**, while a number of linear ridges and valleys along the southwestern limb, such as **Vallis Bouvard**, represent a large-scale impact structure radiating from the Orientale impact basin

Fig. 7.55. Observational drawing of Schickard, a large crater with a multitoned floor near the Moon's southwestern limb. Credit: Grahame Wheatley

further north along the limb. Sunrise takes place from around 10 days to full Moon. Sunset is between around 25 days to new Moon.

Lacus Excellentiae, an irregularly shaped plain of lava centered around the small, flooded crater **Clausius** (25 km), occupies some 22,000 sq km of the lowlands some distance north of Schickard. A number of smaller, darker patches can be seen in the area, notably in the vicinity of **Drebbel B** (16 km) and on the floors of **Drebbel E** (39 km) and **Lehmann E** (37 km). On the whole, the area of Lacus Excellentiae is monotonous, devoid of obvious faulting or features constructed on a grand scale.

Schickard (227 km) is anything but monotonous. One of the largest craters in this part of the Moon, it has low, circular walls and a relatively smooth floor that displays distinct areas of different albedo. For this reason, it is not difficult to locate through binoculars when it is illuminated by a high Sun, even though the crater's rim is not visible. Much of the northern part of Schickard's floor is dark, with several dark, narrow tendrils that proceed south into the gray central plain. A sharply delineated dark area also exists in the southeastern part of the crater floor. Illuminated by a high Sun, a 100 mm telescope will reveal a dozen or so small,

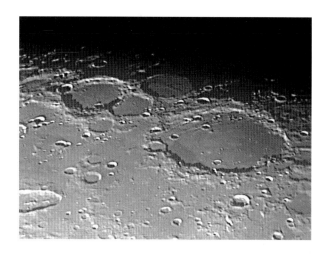

Fig. 7.56. CCD image of Schickard after sunrise. Credit: Peter Grego

bright impact craters that dot the floor, in addition to two larger bright crater rims, **Schickard B** and **Schickard C** (both 12 km), one near the eastern wall and the other near the southwestern wall. A low morning illumination will bring out a considerable amount of topographic detail on Schickard's floor, including **Schickard A** (14 km) near the southern wall, along with a number of prominent craters, grooves and ridges that cross the southwestern part of the floor; these features were produced by secondary impacts from the formation of the Orientale basin, whose main outer rim lies more than 700 km to the northwest. Schickard's northern wall is broken into by the eroded crater **Lehmann** (53 km), and its southwestern flanks are host to **Wargentin A** (21 km).

South of Schickard, **Wargentin** (84 km) appears to be a crater whose floor has been flooded with lava almost up to its rim, producing a dark, circular plateau. Several small wrinkle ridges can be discerned across Wargentin's surface. Illuminated by a low morning Sun, Wargentin appears to lie on the southeastern floor of a larger, considerably eroded (unnamed) crater. **Nasmyth** (77 km), a flooded crater with low walls, adjoins Wargentin's southeastern wall. Nasmyth's southwestern wall is overlain by the larger, more prominent crater **Phocylides** (114 km), whose flooded floor appears noticeably convex when illuminated by a low early morning Sun.

Inghirami (91 km), west of Schickard, is a prominent feature whose floor and ramparts are crossed by linear grooves and ridges. Such extensive radial impact structure emanates from the Orientale impact basin, and it is clearly visible in and around many features of the Moon's southwestern limb. **Vallis Bouvard**, a very large valley 280 km long and around 40 km wide, which extends south of Montes Cordillera, from **Shaler** (48 km) to **Baade** (55 km), is the biggest and best example. To its south, also in the libration zone along the southwestern limb, are the somewhat narrower **Vallis Baade** (160 km long) and **Vallis Inghirami** (140 km long); these are among a number of long radial valleys in the area that make quite a striking spectacle under the right conditions of illumination and libration. The ancient, highly eroded craters **Lagrange** (160 km) and **Piazzi** (101 km), to its south, also display marked linear sculpting originating from the Orientale basin. Patches of darker albedo, visible under a high angle of illumination, also stain the area in and around Piazzi.

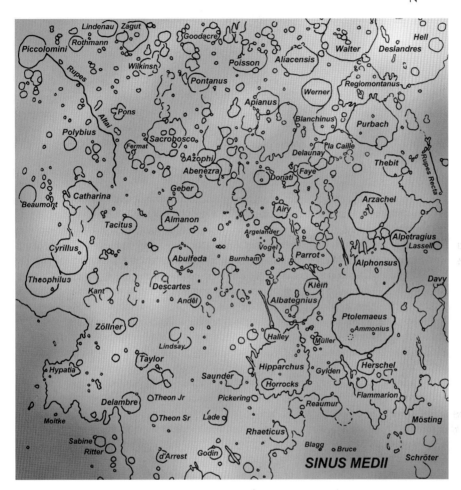

SINUS MEDII

Southeastern Quadrant

Area Thirteen

A collection of some spectacular large craters can be found in the Moon's southeastern central area, the most eye-catching are the magnificent linked trio of **Ptolemaeus**, **Alphonsus** and **Arzachel**. Large craters lie to their east; most notable are the deep-hewn **Albategnius** and eroded **Hipparchus**. Huddled together in the southwestern part of the area, the large craters **Purbach**, **Regiomontanus**, **Aliacensis** and **Blancanus** are a splendid sight when they have emerged from the morning terminator. For all their topographic majesty, few obvious traces of any of

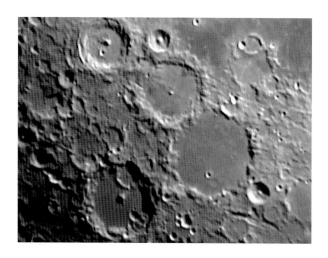

Fig. 7.57. CCD image of the impressive region around Ptolemaeus crater. Credit: Peter Grego

these large craters can be readily discerned under a high angle of illumination. The cratered plains around Albategnius display a number of ridges, grooves and chain craters that are radial to the Imbrian basin. Further east, beyond the **Descartes** highlands, lies another a magnificent linked trio of large craters – **Theophilus**, **Cyrillus** and **Catharina**. To the south, marking the southwestern rim of the impact basin surrounding **Mare Nectaris** (see Area Fifteen), lies the huge, curving scarp of **Rupes Altai**. The sunrise terminator crosses the area from around 4 to 8 days. Sunset is between around 19 and 23 days.

Ptolemaeus (153 km) is an imposing circular plain, sunk some 2,400 m below the level of the highlands, just 9° of latitude south of the center of the lunar disc. Ptolemaeus's walls are markedly striated in an orientation radial to the Imbrian basin far to the north. Illuminated by a high Sun, Ptolemaeus is very difficult to make out clearly, its floor showing up as a mottled gray patch, darker in the west, and peppered with tiny impact craters. All traces of the crater's original central uplift have been submerged. **Ammonius** (9 km), a young, bowl-shaped impact crater on Ptolemaeus's northeastern floor, shows up as a brilliant white spot under a lunar midday Sun, but its surroundings are surprisingly devoid of any prominent ejecta. A low morning or evening Sun will reveal the deep, sharp-rimmed bowl of Ammonius and the low walls of the ghost crater **Ptolemaeus B** (17 km) to its north. Through a 100 mm telescope, the rest of Ptolemaeus's floor can be seen to be dimpled all over with shallow, rimless depressions, older craters that have been flooded with a layer of lava that erupted long after Ptolemaeus was formed. **Herschel** (41 km), a prominent circular crater with broad, internally terraced walls and a group of central peaks, lies immediately north of Ptolemaeus. Further north can be found the disintegrated crater **Spörer** (28 km), and a prominent (unnamed) linear valley, 90 km long and 12 km wide, cuts across its eastern flanks and passes straight through the western wall and floor of **Gylden** (47 km).

Joined to the rough, rubbly southern wall of Ptolemaeus, **Alphonsus** (118 km) is an altogether different looking crater. A solitary central peak, **Alphonsus Alpha** (3,000 m high, with a base diameter of 10 km) rises at the center of a flooded floor that lies some 2,730 m beneath the surrounding highlands; Alphonsus is therefore a few hundred meters deeper than its neighbor, Ptolemaeus. At a low angle of illumination, Alphonsus Alpha appears to protrude through the eastern flanks of a

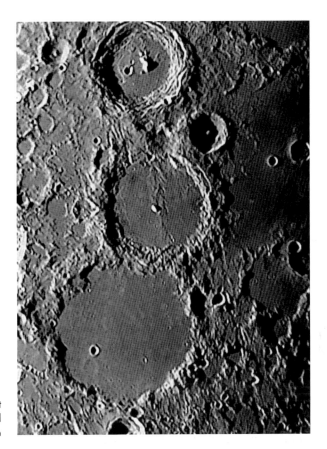

Fig. 7.58. The magnificent craters Ptolemaeus and Alphonsus. Credit: Mike Brown

broad ropy ridge. Other, narrower ridges can be traced across the floor too; these features are aligned radially to the Imbrian basin, and they have been subsequently overlain with a relatively thin veneer of lava. A 150 mm telescope will clearly resolve components of **Rimae Alphonsus**, a network of fine linear rilles that crosses the eastern half of the crater floor; these are best viewed when illuminated by a low evening Sun. A small telescope will easily show three prominent dark, circular patches on Alphonsus's floor: one near the inner western wall, one near the inner northeastern wall and another near the inner southeastern wall. Closer scrutiny with a large instrument will reveal that these dark patches all surround tiny craters that lie along the rilles. These dark-halo craters are in fact volcanic vents surrounded by deposits of dark ash, perhaps formed as early as 3 billion years ago.

Alphonsus's southern wall is disrupted by a large, indistinct (unnamed) crater and several linear valleys. To its southwest, gouged deeply into the highland border of Mare Nubium (see Area Nine), is **Alpetragius** (40 km), a prominent crater with a very large, rounded central massif that measures some 15 km from north to south and rises to a height of around 3,500 m above the crater floor. South of Alphonsus lies the prominent crater **Arzachel** (97 km), similar in size and some topographic features to Copernicus (see Area Five), but considerably older. Arzachel was excavated from an already heavily cratered highland area, but clear traces of external impact sculpting can be seen in its vicinity, especially to the west and south of the crater. Arzachel's inner walls are intricately terraced, and in two places the terracing has produced such pronounced concentric ditches in the wall that they have merited their own official names – **Arzachel E** (31 km long) just inside the southwestern rim, and **Arzachel F** (39 km long), inside Arzachel's southeastern rim. A sizable central peak (19 km long) rises above the center of Arzachel's floor; it is dented by two small craters along its southern flanks. To its east lies the bowl crater **Arzachel A** (10 km), and the eastern part of Arzachel's floor is crossed by the narrow **Rima Arzachel**, a sinuous rille that runs parallel to the inner wall from south to north, best viewed when it is illuminated by a low evening Sun.

Hipparchus (150 km), an ancient crater with considerably eroded walls, lies southeast of Sinus Medii (see Area One). Hipparchus's western wall is jumbled and ill-defined, and parts of the northwestern wall are breached in places and linked directly to Sinus Medii. The eastern wall is more sharply defined, and is cut through by two prominent narrow valleys that are part of the extensive Imbrian impact basin sculpting in the region. Despite its great age, Hipparchus has a generally smooth-looking flooded floor dotted with a few minor craters, low hills and ridges. In the northeast, pressing against Hipparchus's inner wall, the crater **Horrocks** (30 km) has a sharp, though somewhat misshapen, rim and marked internal terracing. On Hipparchus's southern floor can be traced the southern rim of a largely submerged crater, **Hipparchus X** (17 km). **Halley** (36 km), a flat-floored crater with a tiny central hill, intrudes upon Hipparchus's southern flanks, and a narrow gorge (82 km long) proceeds south of Halley, passing just east of the prominent crater **Albategnius** (136 km).

Fig. 7.59. The vast crater Hipparchus. Credit: Mike Brown

Fig. 7.60. The large crater Albategnius. Credit: Mike Brown

Albategnius has a very broad inner wall poc kmarked with innumerable craters, large and small. A linked chain of craters runs north from **Albategnius KA** (7 km), near the base of Albategnius inner northwestern wall, for a distance of 60 km towards Hipparchus. **Klein** (44 km), a well-defined crater with a smooth floor (slightly darker than that of Albategnius) and a small central peak, overlies Albategnius's southwestern wall and intrudes into part of its floor. **Albategnius B**, an oval-shaped crater (16 × 19 km), lies on the northern part of Albategnius's floor, at the base of the inner wall. **Albategnius Alpha**, a slightly west-of-center mountain massif, rises some 1,500 m above Albategnius's smooth, dark floor. Using a 150 mm telescope at high magnification, close examination will reveal a short (northeast–southwest) line of two hills on each side of the main peak, in addition to a tiny indentation on the main peak's summit. Since the peak is composed of uplifted crustal material due to the asteroidal impact that created Albategnius, this tiny crater-like feature is not an indication of a volcanic caldera, but probably a small crevasse at the summit. Albategnius's floor is somewhat dimpled, and peppered with a number of very small craters, like the floor of Ptolemaeus. Under a high angle of illumination, two very faint-light gray linear rays can be traced running west–east across the floor south of the central peak, though their origin is uncertain. Adjoining the southern wall of Albategnius, the disintegrated, irregular shaped **Parrot** (70 km) is cut through by a linear valley, linking it with **Airy** (37 km), a rough-walled crater with a large central mountain. East of Parrot lies **Argelander** (34 km) and the oddly shaped conjoined craters **Vogel** (27 km) and **Vogel A** (21 km), whose alignment suggests that they may have been formed by impacting ejecta from the Imbrian basin. **Burnham** (25 km), to its east, is another strangely shaped crater, a deformed feature with a hilly floor that extends into a small valley to its south.

East of Albategnius, the highlands are a mixture of undulating terrain dotted with medium-sized craters. Under a high Sun, the area has a fairly homogenous gray tone dotted with a few small, bright impact craters and overlain with traces of Tycho's rays. The deep, sharp-rimmed bowl crater **Hipparchus C** (17 km) lies at the center of the region's most prominent ray system, which spreads uniformly to a distance of around 50 km from its rim; but it is by no means a spectacular splash. Southeast of Hipparchus, the craters Halley, **Hind** (29 km), Hipparchus C and **Hipparchus L** (11 km) form a prominent line of craters of decreasing size. To their northeast, low, undulating plains and eroded craters, like **Saunder** (45 km) and **Lade** (56 km), lie between mountainous ridges. In the far northeast, the deep, bowl craters **Theon Senior** (18 km) and **Theon Junior** (19 km), and **Delambre** (52 km), a crater with prominent, terraced walls, lie beyond the southwestern border of Mare Tranquillitatis (see Area One).

Some 250 km east of Albategnius, the highly eroded crater **Descartes** (48 km) has an obliterated northern wall, and its floor is circled by the remnants of an inner ring. The crater's southwestern rim is overlain by the deep, bright bowl of **Descartes A** (15 km), and a prominent bright patch of ray material from **Descartes**

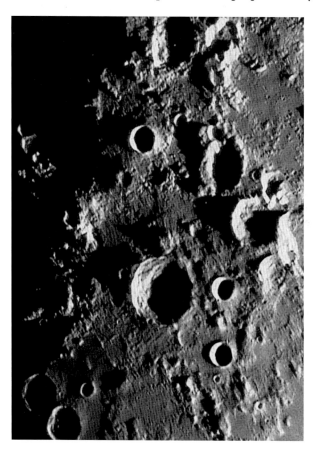

Fig. 7.61. Region around Delambre. Credit: Mike Brown

E (5 km) spreads across Descartes' northeastern flanks. A great number of ancient eroded craters can be found in the highlands surrounding Descartes. In April 1972, Apollo 16 landed on the eastern floor of one of these ancient craters, an unnamed feature 42 km across, 60 km north of Descartes itself. **Abulfeda** (62 km), a circular crater with broad-terraced inner walls and a smooth-flooded floor, lies a short distance southwest of Descartes, and is the most prominent topographic feature in the Descartes highlands.

Proceeding south of Abulfeda, there is a curving alignment of sharp-rimmed craters with broad inner walls – **Almanon** (49 km), **Geber** (45 km), the conjoined trio of **Abenezra** (42 km), **Abenezra C** (42 km) and **Azophi** (48 km), and **Playfair** (48 km). Their chance alignment lends the illusion that they mark part of the southeastern margin of a vast circular plain, 250 km in diameter, its western border marked by Vogel, Argelander and Airy, and whose center is marked by the small, bright ray crater **Abulfeda E** (3 km). **Sacrobosco** (98 km), southeast of Azophi, is an eroded crater with a low, dented rim and a broad, ragged inner wall surrounding three sizable craters that lie on its floor. Playfair and the smooth-floored **Apianus** (63 km) overlie the wall of **Playfair G** (115 km), and the flooded **Krustenstern** (47 km) forms a deep bay to its south. **Poisson**, some distance south of Apianus, is an irregularly shaped multiple crater extending east–west (44×69 km). A group of larger, more prominent and better-preserved craters lies to the west, beginning with **Aliacensis** (80 km), a deep feature with broad inner walls and a smooth floor overlooked by a small, isolated peak lyinf slightly north of center. Bright ray material from Tycho dusts the northern part of the crater's floor, and its eastern rim bulges markedly to the east. **Werner** (70 km) almost touches the north-western rim of Aliacensis. It has exceptionally well-ordered inner terraced walls that surround a smooth floor lying 4,200 m beneath the level of its sharp rim, and a group of three prominent mountain peaks protrudes from the crater's floor. The eroded crater **Blanchinus** (63 km) links Werner with **la Caille** (68 km) to the north, and a remarkable double crater, **Delaunay** (46 km), lies beyond la Caille's north-eastern rim. Linear rays from Tycho cross la Caille's smooth floor, and these can be traced across the broad floor of **Purbach** (118 km) to the southwest. A group of hills on Purbach's western floor outlines part of the rim of a flooded crater, **Purbach W** (25 km). Purbach's western rim is dented with craters, including the teardrop-shaped **Purbach G** (37 km) in the northwest, and a chain of craters, **Purbach M** (15 km), **Purbach L** (19 km) and **Purbach H** (24 km), which stretch from Purbach along the southeastern border of Thebit P (see Area Nine). An elongated, somewhat irregularly shaped and considerably eroded crater, **Regiomontanus** (126 km wide), joins Purbach's southern wall. Regiomontanus's western wall is in an advanced state of disintegration, but its eastern wall is rather more clearly defined. Of note is a mountainous headland that projects from the crater's inner northern wall beyond the center of the floor, upon whose summit lies the crater **Regiomontanus A** (6 km), probably an impact crater on a central uplift rather than a caldera on a lunar volcano.

Theophilus (100 km), **Cyrillus** (98 km) and **Catharina** (100 km), a linked trio of craters on the western border of Mare Nectaris, form one of the Moon's most well-known landmarks, a superb sight through any telescope when illuminated by an early morning or late afternoon Sun. Theophilus is an imposing structure. Its prominent circular rim rises 1,200 m above the level of Mare Nectaris, to its east. Inside, broad walls descend in an intricate series of terraces some 4,400 m to an impact melt floor. At its center, a group of three hefty central peaks – **Theophilus**

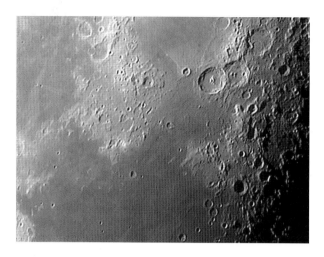

Fig. 7.62. Theophilus, Cyrillus and Catharina, and the area to their northeast – Sinus Asperitatis and mare Tranquillitiatis. Credit: Brian Jeffrey

Alpha, Theophilus Phi and **Theophilus Psi** – rise to heights of 1,400 m above the floor. Low foothills link Theophilus Alpha and Theophilus Psi to the inner southeastern and northwestern walls, respectively, but the northeastern part of Theophilus's floor is uniformly smooth. **Theophilus B** (9 km), a small bowl crater, lies amid the terraces inside the crater's northwestern rim. Theophilus' outer ramparts display an extensive system of radial ridges, grooves, secondary impact craters and crater chains that spread east across the plains of Mare Nectaris and north across Sinus Asperitatis to distances exceeding 100 km. Under a high illumination, all traces of radial impact structure disappear, Theophilus' rim appearing as a ghostly gray ring enclosing the brilliant spots of its central peaks and Theophilus B.

Theophilus overlies the northeastern wall of Cyrillus, a similar sized crater that shares many of the same topographic features, only it is considerably more ancient and eroded than its neighbor. Cyrillus's ramparts are lower and less orderly than

Fig. 7.63. CCD image of the linked trio of craters Theophilus, Cyrillus and Catharina. Credit: Peter Grego

those of Theophilus, but traces of its original external impact sculpting can be found to its north, notably in the radial grooves that cut through **Ibn Rushd** (33 km) to its northwest. Cyrillus's rough inner southwestern wall is host to **Cyrillus A** (15 km). Three rounded mountains – **Cyrillus Alpha**, **Cyrillus Delta** and **Cyrillus Eta** – rise to heights of 1,000 m above Theophilus's floor, slightly northeast of center. A rille (25 km long) curves southwest from Cyrillus Alpha across the crater floor to just east of Cyrillus A, and a ridge along Cyrillus's inner eastern wall creates the illusion that there is another curving rille when the area is illuminated by a late afternoon Sun.

A rough mass of mountain ridges breaks from Cyrillus's southern wall and runs to the south across the northeastern ramparts of Catharina. Under a low angle of illumination, Cyrillus appears connected to Catharina by a broad valley. Catharina has a low, eroded rim that is dented by a number of craters, notably, in the northeast, by a small chain of craters that runs south from **Catharina B** (19 km). Much of Catharina's northern floor is occupied by the flooded **Catharina P** (49 km), parts of whose southern edge are breached and level with the rest of Catharina's floor. **Catharina S** (14 km), the rim of another submerged crater, lies on Catharina's southern floor, touching the inner wall.

Under a low evening Sun, the ridges east of Catharina can be traced around to Fracastorius, a bay crater in southern Mare Nectaris (see Area Fifteen), 275 km away. These ridges are part of the highly eroded inner mountain ring of the Nectaris impact basin. Another eroded mountain ring (originally 660 km in diameter) concentric to the Nectaris multiring impact basin can be traced from Catharina's southern ramparts to the south, between Fracastorius and **Piccolomini** (88 km). Piccolomini is a prominent, strongly internally terraced crater with a very large cluster of central peaks that reach heights of 2,000 m. By far the most prominent section of the Nectaris multiring impact basin – the huge east-facing scarp face of **Rupes Altai** – runs north of Piccolomini around to the west of Catharina, creating a giant fault that cuts through the mountains for a distance of around 500 km. East of Rupes Altai, the lunar crust has dropped more than 1,000 m below the level of the scarp's rim. Although Rupes Altai hardly cuts a perfect arc, it shows

Fig. 7.64. Rupes Altai, a huge arcuate fault delineating the southwestern quadrant of the larger Nectaris basin. Credit: Peter Grego

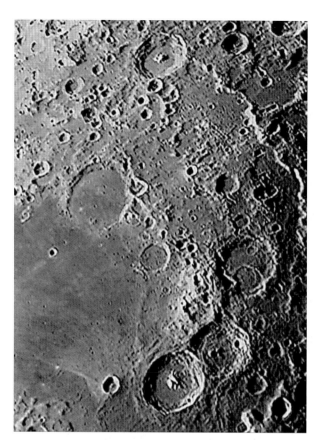

Fig. 7.65. CCD image of the linked trio of craters Theophilus, Cyrillus and Catharina, with Rupes Altai. Credit: Mike Brown

up through binoculars and small telescopes as a gleaming curve to the southwest of Mare Nectaris when it is illuminated by the morning Sun, and as a broad black swath under a low evening Sun. Measuring some 880 km in diameter, the remaining sections of the Nectaris basin's outer ring are more difficult to observe, but parts of them can be traced from the mountains north of Theophilus east of Hypatia, south to the broad tongue of **Mons Penck** (30 km wide, 4,000 m high) west of Theophilus.

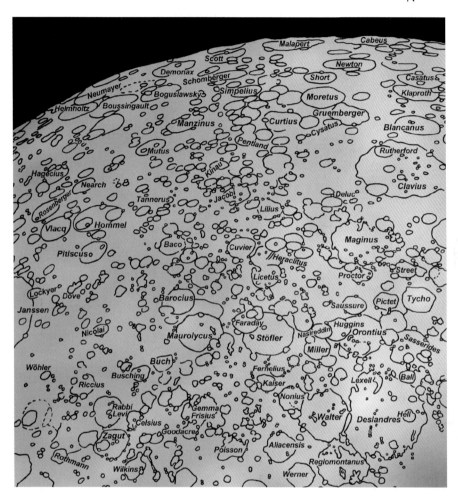

Area Fourteen

This heavily cratered, mareless region of the eastern part of the Moon's southern uplands includes all the major features that lie in the vicinity of the south pole. Much of the northern part of this area is covered with easily discernible rays emanating from Tycho to the west (see Area Ten). Prominent large craters include **Walter** in the northwest, **Stöfler** and **Maurolycus** to its southeast, **Hommel** in the east and **Moretus** in the south. Sunrise takes place from around 4 to 8 days. Sunset is between around 19 and 23 days.

   Walter (120 km), a prominent crater with terraced inner walls, indents the eastern wall of the larger Deslandres (see Area Ten). Much of Walter's floor is smooth and flooded, but a single ghost crater, **Walter E** (12 km), can be found near its inner western wall, and a cluster of five similar-sized craters in various states of erosion, including the sharp-rimmed **Walter A** (10 km), lie on top of a hilly area on Walter's northeastern floor. Under a high Sun, several rays from Tycho can clearly be discerned on Walter's floor. Several craters lie along Walter's eastern rim, including the overlapping pair of **Walter L** (25 km) and **Walter K** (18 km): the pair also lie on the northern rim of the highly eroded, slightly elongated crater **Nonius** (62 × 57 km), whose northern floor is crossed by two very large bulging ridges. To the southeast, two adjoining smooth-floored craters, **Kaiser** (52 km) and **Fernelius** (65 km), lead to **Stöfler** (126 km). The entire western half of Stöfler's floor is smooth and dusted with bright rays from Tycho, and its western rim is dented by the deep bowls of **Stöfler K** (18 km) and **Stöfler F** (18 km). A prominent, curving mountain ridge – possibly part of an ancient crater rim – crosses its eastern floor. **Faraday** (70 km) overlies Stöfler's eastern wall, and Faraday's own wall is heavily covered by craters, notably in the south, by a line of four overlapping craters of steadily decreasing size, from **Stöfler P** (32 km) eastwards.

East of Faraday lies the imposing crater **Maurolycus** (114 km), whose broad inner eastern wall displays intricate terracing, though its inner western wall is somewhat rougher and more disordered. The northern part of Maurolycus's floor is hilly and peppered with small craters, and a ragged group of mountains cut through by a broad and deep curving trench lies just north of center. The largest peak, just east of center, is a markedly pyramidal structure with faceted sides

Fig. 7.66. Stöfler region.
Credit: Mike Brown

capped by a tiny crater, the latter feature visible only through large telescopes. Maurolycus's floor is markedly convex, and its eastern floor slopes down considerably to meet its inner eastern wall. When the crater is illuminated by a low evening Sun, the eastern part of the floor is immersed into deep shadow, while parts of the floor further west remain brightly illuminated. Maurolycus overlies most of an equally large (unnamed) crater to its south, whose broad, well-structured southern wall is well preserved. Southeast of Maurolycus, **Barocius** (82 km) has a breached northern wall, overlain by **Barocius B** (41 km). An irregular cluster of low mountains lies at its center, and on Barocius' southwestern floor is the flooded crater **Barocius W** (15 km).

The walls of **Gemma Frisius** (88 km), some distance north of Maurolycus, have been extensively pummeled by large impacts, giving it the appearance of a bear's pawprint. **Goodacre** (46 km), in the northeast, is the largest of these superimposed craters. Both Gemma Frisius and Goodacre have small, single central mountains, Gemma Frisius's lying just northwest of center. Proceeding east from Gemma Frisius's rim, a linked line of low-walled, smooth-floored craters of decreasing size includes **Gemma Frisius A** (46 km), **Gemma Frisius B** (41 km) and **Gemma Frisius C** (37 km). Further east, the prominent duo of **Zagut** (84 km) and **Rabbi Levi** (81 km) adjoin the prominent western wall of a large and otherwise considerably eroded unnamed crater, elongated east–west (105 × 130 km). The northwestern floor of this unnamed crater is host to **Lindenau** (53 km), a prominent crater with a double-rimmed western wall and five sizable central mountains arranged in a

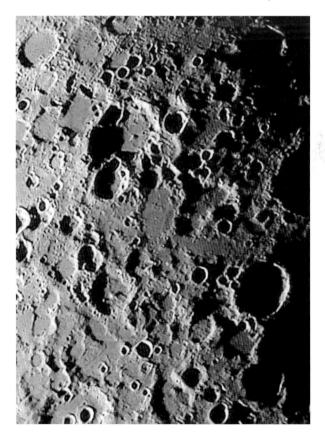

Fig. 7.67. Region around Gemma Frisius. Credit: Mike Brown

starfish pattern. Southeast of Rabbi Levi lies a concentrated area containing dozens of sizable young impact craters. Among them, **Riccius** (71 km) probably holds the record for being the Moon's most heavily cratered crater; only its western wall has escaped bombardment. The rest of Riccius has been completely obliterated by more than a dozen craters larger than 8 km across. Further east (100 km) **Stiborius** (44 km) is a deep, prominent crater with a central mountain massif and a highly deformed northeastern wall that has collapsed in a series of major landslides. Stiborius lies near the center of a large, eroded unnamed crater (93 km), visible only at low angles of illumination. Clear signs of radial sculpting from the Nectaris impact basin can also be found in this area (see Area Sixteen).

Heraclitus (90 km) has a prominent central mountain spine (30 km long) oriented southwest–northeast. Heraclitus's southern floor is indented by **Heraclitus D** (49 km), its northern rim is breached by **Licetus** (75 km), and **Cuvier** (75 km) indents its eastern rim. A linked line of large craters runs to the east, from **Clairaut** (75 km), easily identifiable by its two prominent craters on its floor, through **Clairaut A** (35 km) and **Baco B** (40 km) to **Baco** (70 km). Further east, 150 km by the walls and floor of the large, eroded crater **Hommel** (125 km) are overlain by sizable craters, the largest of which are **Hommel A** (50 km) in the northeast and **Hommel C** (47 km) in the northwest. On Hommel's floor, between these two superimposed craters, a U-shaped mountain ridge marks Hommel's eroded central uplift. **Hommel H** (33 km), to the north, links Hommel with **Pitiscus** (82 km), a sharp-rimmed crater with a 12-km-long central mountain at whose eastern end lies the deep bowl crater **Pitiscus A** (9 km). Secondary impact structure can be seen in the relatively smooth areas immediately surrounding Pitiscus. Adjoining Hommel's eastern wall is **Vlacq** (89 km), a large crater with a sharp eastern rim and a somewhat eroded western wall. At its center lies a prominent mountain spine 19 km long; sunk into the crater floor at the far northern end of the mountain is a small unnamed crater (5 km), and to its west lies another unnamed sunken crater, somewhat oval-shaped (6 × 11 km). A 200 mm telescope may reveal the tiny crater that lies just west of center on the floor of the latter feature. An unnamed sunken crater (13 km) lies on the western floor of **Rosenberger** (96 km), a prominent crater that adjoins Vlacq's southern wall. Numerous smaller craters can be discerned near the center of Rosenberger's floor through a 150mm telescope. **Rosenberger D** (47 km), adjoining Rosenberger's southern wall, is noteworthy for its smooth-sloped, elongated central hill, which joins with its inner northern wall. To its west lies **Nearch** (76 km), a deep, smooth -floored crater. A little further towards the southeastern limb is the somewhat more battered **Hagecius** (76 km), whose southern wall is superimposed on by several sizable craters. To the north, **Hagecius A** (65 km) is an older crater, half of which is neatly overlain by the larger Hagecius.

South of Heraclitus, another prominent, linked line of craters runs east from **Lilius C** (34 km) through **Lilius** (61 km) and **Lilius A** (37 km) to **Jacobi** (68 km), and beyond to several smaller, overlapping craters. Lilius is easily recognizable because of the prominent pyramidal central peak that overlooks its smooth floor, while Jacobi's floor is dented with a cluster of five large craters. Between Jacobi and the southeastern limb is the prominent, close duo of **Manzinus** (98 km), a deep crater with a smooth, noticeably convex floor, and **Mutus** (78 km), with two sizable craters on its eastern rim and one on its western floor.

Manzinus and Mutus can be used as stepping stones to identify features along the southeastern limb and the libration zone. Southeast of Mutus, the large crater **Boussingault** (131 km) is an impressive feature when viewed at a favorable libration under a mid-morning or late afternoon Sun. **Boussingault A** (79 km) occupies its entire northern floor, and its floor lies at least 5,000 m beneath the rim of Boussingault. Under a low angle of illumination, Boussingault appears to have a substantial double wall. **Boussingault E** (103 km) adjoins Boussingault's northwestern wall, the deep polygonal crater **Boussingault K** (27 km) marking their meeting place. Boussingault E has a chaotic mountainous floor, and its northern rim is overlapped by **Boussingault B** (60 km) and **Boussingault C** (21 km). On the relatively smooth plain to the north, beginning immediately west of **Boussingault T** (19 km), a distinct but eroded chain of craters runs north for 60 km, terminating south of **Nearch A** (49 km). This chain of craters is likely to be a secondary impact feature, but the crater that it emanated from may be Boussingault itself, to the south, or Vlacq to the north. In the libration zone beyond Boussingault lie **Helmholtz** (95 km) and **Neumayer** (76 km). The center of **Hale** (84 km), a deep crater with complex, terraced inner walls and a large central peak system, lies just beyond the 90°E line of longitude, some 100 km east of Neumayer. **Boguslawsky** (97 km), a crater with broad inner walls and a smooth floor, almost touches the western wall of Boussingault. In the libration zone to its south lies **Demonax** (114 km), an older crater with somewhat eroded walls and a compact group of central peaks. Its southern rim lies just 10° (300 km) from the Moon's south pole.

The Moon's heavily cratered south polar region can be a confusing place to visit telescopically: not only do the craters lie cheek to jowl and are highly foreshortened, but the effects of libration must be contended with. Regular visits to the region, along with attempts to identify certain key features, will enable the observer to navigate the area successfully. **Moretus** (114 km), a deep crater with terraced inner walls and a large central massif rising above a smooth floor, is one of the most prominent landmarks in the polar region. Moretus' central mountain casts a long pointed, shadow that touches its inner walls when illuminated by a low Sun. To the northwest lies the crater **Gruemberger** (94 km), its southern wall heavily disrupted by the impact-sculpted flanks of Moretus. A little distance northeast of Moretus lies **Curtius** (95 km), the remnants of its central uplift barely protruding from the western part of its flooded floor. A large, irregular, unnamed crater lies east of Moretus, and is overlain in the east by **Simpelius C** (42 km) and **Simpelius D** (51 km). **Simpelius** (70 km), a deep crater with broad inner walls and a group of low central hills, adjoins the wall of Simpelius D. Towards the limb lies **Schomberger** (85 km), a prominent crater with a sharp rim and a large central massif, and the somewhat more disintegrated crater **Scott** (108 km) to its south, in the libration zone. Beyond Scott, lying on the 90°E line of longitude, **Amundsen**

Fig. 7.68. Observational drawing of the Moon's south polar region This region presents one of the Moon's greatest observing challenges since the features here all appear extremely foreshortened, and its appearance changes considerably owing to illumination and libration effects. Credit: Grahame Wheatley

Fig. 7.69. The large crater Moretus is a prominent land-mark near the southern limb. Credit: Mike Brown

(105 km) is the largest crater near the lunar south pole, its southern rim lying just 100 km from it. Amundsen is a deep crater with well-preserved internal terracing and a large group of central mountains.

Almost touching the southern wall of Moretus, **Short** (71 km) is sunk into the uplands, its rim almost level with the surrounding terrain. A small, unnamed crater (6 km) lies at the center of its floor, and Short's southeastern rim is intruded upon by **Short B** (51 km), which, in turn, joins the northwestern wall of **Short A** (37 km). Short A lies on the Moon's central meridian, 13° (390 km) from the south pole. Immediately southwest of Short is **Newton** (79 km), which covers much of the northwestern part of a large, unnamed crater (83 km) whose southern wall is over-lain by **Newton A** (63 km). Further south, in the libration zone on the southern limb, the eroded crater **Cabeus** (98 km) lies around 100 km from the south pole. The south pole itself can never be observed directly; even at the most favorable libration and angle of illumination, it is obscured by higher ground and lies in constant shadow.

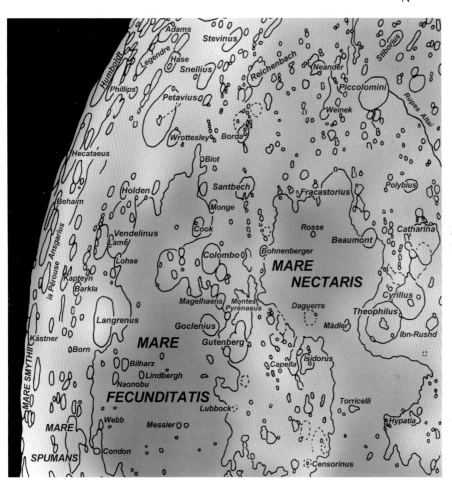

Area Fifteen

The gray, undulating plain of **Sinus Asperitatis** (Bay of Asperity) links the southern reaches of Mare Tranquillitatis to the north of **Mare Nectaris** (Sea of Nectar). Flooded **Fracastorius** makes a neat, semicircular bay on the southern coast of Mare Nectaris. Beyond the small patch of **Mare Spumans** (Foaming Sea), in the libration zone of the eastern limb, lies **Mare Smythii** (Smyth's Sea). Towards the eastern limb, the southern half of **Mare Fecunditatis** (Sea of Fertility) flows into a large, irregular bay west of the crater **Vendelinus**. The prominent bright crater **Langrenus** lies on Mare Fecunditatis's eastern coast, some distance north of Vendelinus. Equidistant from Vendelinus, to its south, lies mighty **Petavius**, a crater with a sprawling central massif and large rilles on its floor. Numerous bright rays are traceable across this part of the Moon. In western Mare Fecunditatis, the double linear ray of **Messier** is a delight to view. Asymmetric rays emanate from **Mädler** in Sinus Asperitatis and **Petavius B**. Rays from Tycho (see Area Ten) can be traced across the western part of the area, while rays from **Stevinus A** and **Furnerius A** spread from the south (see Area Sixteen). Sunrise is between new Moon to around 4 days. Sunset takes place after full Moon to around 19 days.

An unusual, teardrop-shaped crater, **Torricelli** (23 × 30 km), lies on the north-eastern floor of a large flooded crater, **Torricelli R** (81 km), whose northwestern rim has been completely submerged by the lava of northern Sinus Asperitatis. Low wrinkle ridges can be traced to Mare Tranquillitatis, to the north, while, to the south, prominent radial ridges from Theophilus (see Area Thirteen) scar the mare. A little distance east of Theophilus, **Mädler** (28 km), an elongated crater whose rim is intruded upon by a prominent 30-km-long ridge to the north, has a bright, single ray that stretches for 130 km eastwards across northern Mare Nectaris, covering the southern floors of two ghost craters, an unnamed one (54 km) and **Daguerre** (46 km), both of whose southern rims have been completely submerged. The highlands to the north of Mare Nectaris are host to **Capella** (49 km), a battered crater with a large, prominent central mountain, and **Isidorus** (42 km), an older but better-preserved feature with a single, deep keyhole-shaped crater, **Isidorus A** (11 km) on the western part of its smooth floor. Capella is crossed by **Vallis Capella** (110 km long), a coarse valley made up of a chain of sizable craters of diminishing size that runs from the north of Capella to its southeast, across its eastern floor. West of Isidorus, the western side of a plateau (25 km wide) presents a curving scarp face several hundred meters high overlooking a smooth section of Sinus Asperitatis' shoreline. **Gaudibert** (33 km), along the coast of Mare Nectaris to the southeast, is one of a number of inconspicuous, somewhat irregular-shaped craters in the region whose floors are intricately sculpted with ridges. Other examples include **Isidorus B** (41 km), **Censorinus C** (28 km) and **Maskelyne A** (27 km), all some distance to the north. Just west of Maskelyne A lies the tiny crater **Censorinus** (3.8 km), conspicuous because of its brilliant (25-km-diameter) collar of ejecta.

Rimae Gutenberg, a group of parallel linear rilles, cuts through the highlands from Censorinus C across the floor of **Gutenberg** (74 km), 350 km to the southeast. A 100mm telescope will show the rille within Gutenberg, but the whole system requires a 150mm telescope to resolve adequately. Gutenberg lies on the western coast of Mare Fecunditatis. Its eastern wall is overlain by **Gutenberg E** (31 km), and a breach in its wall has allowed the dark lava flows of the mare to enter both craters. A group of peaks arranged in a semicircle on Gutenberg's floor likely represents a combination of the crater's original central uplift and the rim of a submerged crater. Its peaks in the north are intercepted, but not cut through, by the main Gutenberg rille, which runs northwest to southeast across the crater's floor. Gutenberg's southern wall overlies and extends southwards into **Gutenberg C** (43 km) in a number of sharp ridges. Gutenberg and its adjoining craters take on an odd lobster pincer appearance when illuminated by an early morning Sun. **Montes Pyrenaeus**, a winding strip of mountains 250 km long, rise to the south of Gutenberg.

Mare Fecunditatis, a collection of completely flooded craters and impact basins, is an extensive dark mare towards the eastern limb of the Moon. It has an irregular outline, measuring around 900 km from north to south, 600 km at its widest in the north, with a total surface area of around 326,000 sq km. Illuminated by a high Sun, rays from a number of impact craters can be clearly traced across various parts of the mare, including a single ray from Tycho, some 1,800 km to the south-west. Under a low angle of illumination, prominent wrinkle ridges show up in the northern and eastern parts of the mare. **Dorsa Cato** (140 km long), **Dorsum Cushman** (80 km long) and **Dorsum Cayeux** (130 km long) extend south across the mare from Taruntius (see Area Three) in the north of Mare Fecunditatis. In the eastern central part of the mare, **Dorsa Geikie** (240 km long) extend south to meet the broad main ridge of **Dorsa Mawson** (180 km long). Additionally, many completely submerged craters can be discerned over the mare, their presence indicated by very low, circular ridges. One example is **Goclenius U** (20 km), which lies beyond the western tip of Dorsa Mawson.

Messier (7 × 13 km) and **Messier A** (14 × 9 km), two neighboring craters in northwestern Mare Fecunditatis, are fascinating to observe. Messier is a deep, oval crater whose long axis lies east–west, while Messier A is a deep, circular bowl with a raised semicircular lip beyond its western rim. The pair are separated by just 6 km. Messier lies along a broad ray that extends from Taruntius in northern Mare Fecunditatis, tapering to a point 100 km south of Messier. Close examination will reveal a fine butterfly pattern of rays spreading in a narrow fan to Messier's north and south, superimposed upon the larger ray from Taruntius. Extending from the west of Messier A is one of the Moon's most remarkable ray systems, a close pair of linear rays that reach the western border of the mare 150 km away, diverging very slightly over their course. It is likely that Messier and Messier A were produced when a small asteroid hit the lunar surface from the east at a very shallow angle, probably less than 5°. Messier was formed from the primary impact, and Messier A was carved out by the impact of a large fragment of this asteroid that rebounded and careered into the Moon a little further downrange. This scenario would explain the elongated shapes of the pair and their unique pattern of ejecta. **Rima Messier** (94 km long), a very narrow linear rille visible through a 200mm telescope, cuts across the mare from around 30 km northwest of Messier A to the coast near Montes Secchi (see Area Three).

In western Mare Fecunditatis, the slightly irregular shaped crater **Goclenius** (51 × 62 km) is crossed by a single linear rille, and, as with Gutenberg to its north-west, this rille intersects but does not cut through the crater's central mountain, nor does it appear to slice into the crater's walls. An extension of this rille, one of components of the **Rimae Goclenius** system, extends northwest of the crater wall, intercepting the wall of Gutenberg E and picking up again to its north, covering a distance of 150 km. Another linear rille (145 km long) can be found to its east, running from **Goclenius B** (8 km) to the hills south of the small, flooded crater **Lubbock** (14 km). These features were produced from the same set of crustal stresses that gave rise to Rimae Gutenberg in the hills to their northwest.

South of Goclenius, the dark-floored **Magelhaens** (41 km) adjoins **Magelhaens A** (33 km), on the northern flanks of the prominent conjoined duo **Colombo** (76 km) and **Colombo A** (37 km). Colombo has a particularly broad and complex inner eastern wall and a small group of central peaks, while Colombo A is flooded and resembles Magelhaens. The highlands east of Colombo are riddled with numerous low-walled flooded craters.

Beyond the crater **Webb** (22 km), on the northeastern coast of Mare Fecunditatis, lies **Mare Spumans**, a small, irregular patch of mare measuring 100 km wide × 150 km long, with a surface area of 16,000 sq km. Its borders are indented with numerous bays like **Webb C** (38 km) on its western shore, and flooded craters like **Pomortsev** (23 km) on its eastern shore. Dozens of similar features abound to the east and northeast, in and around Mare Spumans and Mare Undarum (see Area Three). On the eastern limb in the libration zone, **Mare Smythii** (200 km in diameter) is a roughly circular sea. Although the whole of Mare Smythii is presented towards the Earth during a favorable libration, its near-limb position makes it appear very foreshortened. It can disappear beyond the eastern limb altogether during an unfavorable libration. Mare Smythii contains many flooded craters, like the conjoined duo of **Kiess** (63 km) and **Widmanstätten** (46 km) on its southern shore. To its southwest lie the large craters **Gilbert** (107 km) and **Kästner** (105 km), both with low walls and undulating floors.

Langrenus (132 km) is a spectacular crater on the eastern shore of Mare Fecunditatis. Unique among large craters of its size and age, its exceedingly complex, terraced inner walls are remarkably free from any sizable superimposed craters. Langrenus's floor is overseen by a pair of large central mountains, **Langrenus Alpha** and **Langrenus Beta**, both of which are higher than 1,500 m. The northern part of the crater floor is noticeably rougher than the southern part, the difference in texture becoming obvious under a low angle of illumination when viewed through a 100 mm telescope. A mass of radial ridges spreads across the mare to the north and west of Langrenus. Secondary impact craters dot the landscape – notable are clusters of small craters near **Langrenus DA** (4 km), **Al-Marrakushi** (8 km) and a linear group near **Langrenus V** (3 km). These are resolvable in a 150mm telescope under ideal conditions. Another cluster of small, secondary craters poc kmarks the terrain north of the smooth-floored crater **Bilharz** (43 km). **Atwood** (29 km) adjoins Bilharz's eastern wall, and **Naonubu** (35 km) lies immediately north of Atwood. The three craters make a prominent group to the northwest of Langrenus. Under a high Sun, Langrenus appears as a bright, circular patch, its inner walls and central peaks appearing slightly brighter than its floor. Langrenus's rays can be traced in all directions across the surrounding landscape – particularly across Mare Fecunditatis, where they show up better against the dark terrain – for distances of more than 250 km. East of Langrenus, a line of craters of increasing size – **Barkla** (43 km), **Kapteyn** (49 km) and **la Pérouse** (78 km) – proceeds towards the limb. Each has a sharp rim, internal terracing and a deep floor containing a central mountain.

Vendelinus (147 km), on Mare Fecunditatis' southeastern shore, is an ancient crater with eroded, heavily cratered ramparts. Rays from Langrenus, 150 km to the north, can be traced across its relatively flat floor. **Holden** (47 km), a deep crater with terraced walls, adjoins Vendelinus's southern wall. **Lohse** (42 km) indents Vendelinus' northwestern rim, and its northeastern wall is overlain by **Lamé** (84 km). A chain of seven large craters proceeds south along the inner wall of Lamé through **Lamé G** (21 km), on Lamé's southeastern rim, to **Lamé P** (16 km). Towards the southeastern lunar limb can be found an unnamed irregular-shaped, smooth gray plain, its southern border extending into the broad semicircular bay of **Balmer** (112 km), a crater whose northern rim has been completely submerged. In the libration zone further east lie the central peak craters **Behaim** (55 km) and **Gibbs** (77 km). **Hecataeus** (127 km) lies further south along the limb, its northern wall overlapping the smaller crater **Hecataeus K** (90 km).

Petavius (177 km), a magnificent formation, lies a short distance beyond the southern shore of Mare Fecunditatis. Its walls are exceedingly complex, displaying extensive terracing and deep grooving. A sprawling mountain massif with a base 36 km wide rises in several sizable individual peaks – a 2 km crater indents the summit of the southwestern one. A mixture of several linear and sinuous rilles,

Rimae Petavius, cut across Petavius's undulating floor. Observable through a 60 mm telescope, the largest of the Rimae Petavius cuts a straight path for 50 km, from the central mountains to a deep ditch running along the base of the inner southwestern wall. Another linear rille runs a short distance across the floor to the south of the central mountains to near **Petavius A** (5 km). A large, irregular depression adjoining the northern central mountains appears to be the origin of a sinuous rille (45 km long) that meanders to the base of the inner northern wall. Much of Petavius's eastern floor is crossed by a narrower sinuous rille (95 km long) that originates in the crater's southeastern wall and peters out in the northeastern part of the floor. The latter feature is resolvable through a 150mm telescope. Petavius's outer flanks display intricate radial impact sculpting and several large crater chains, particularly visible to the north of the crater. Further north, on the coast of Mare Fecunditatis, the prominent crater **Petavius B** (35 km) is the center of a bright ray system that spreads in a butterfly-wing pattern to the east and west.

Wrottesley (57 km), a deep crater with broad, terraced inner walls and a small central massif, adjoins Petavius's northwestern rim. On the opposite side of the crater, just outside Petavius's southeastern rim, lies the crater **Palitzsch** (41 km), whose northern wall extends northwards into **Vallis Palitzsch** (110 km long, averaging 19 km wide), a series of large, connected craters that lie around the eastern edge of Petavius. South of Palitzsch, the disintegrated crater **Hase** (83 km) is overlain by **Hase D** (54 km). **Snellius** (83 km), a deep but considerably battered crater, lies southwest of Petavius. It is crossed by **Vallis Snellius** (500 km long), one of the Moon's lengthiest valleys, made up of a connected chain of mainly shallow craters that extend eastwards from the crater **Borda** (44 km) to the south of Hase D.

Humboldt (207 km), a huge crater in the libration zone east of Petavius, appears most impressive when illuminated by a late afternoon Sun at a favorable libration. A line of mountains (100 km long) crosses the northeastern section of Humboldt's floor, north of which lies a deep bowl crater (20 km). Dark patches lie on the floor along Humboldt's inner western wall and on the northeastern floor. An extensive system of linear rilles crosses the southern part of the floor, and another large linear rille cuts across the far-northern floor. Humboldt slightly overlies the southern wall of the older Hecataeus, and the eroded crater **Barnard** (98 km) just overlies Humboldt's southeastern wall.

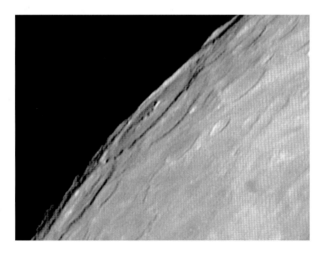

Fig. 7.70. Humboldt, a huge crater near the southeastern limb. Credit: Brian Jeffrey

Mare Nectaris is a square-shouldered sea, around 350 km wide. The multiring basin in which Mare Nectaris lies is far more extensive, an outer ring of which measures around 880 km in diameter, marked by Rupes Altai in the southwest (see Area Thirteen). Gray rays from Theophilus (see Area Thirteen) on the mare's northwestern coast can be traced across much of the mare, in addition to the brighter rays of Mädler in the north and linear rays from Tycho, 1,280 km to the southwest. A broad ridge 115 km long connects the southeastern flanks of Theophilus with **Beaumont** (53 km) on the southwestern shore of Mare Nectaris. Beaumont's floor has a smattering of low hills and small craters; a narrow rille (10 km long) lies on the floor to the west of a small gap in Beaumont's eastern wall. A considerably larger gap exists in the northern wall of **Fracastorius** (124 km), a flooded crater that forms a deep bay on the southern coastline of Mare Nectaris. Traces of its largely obliterated northern rim can be discerned in several low hills and ridges. A narrow rille (70 km long) cuts across the southern half of Fracastorius' floor from west to east, intercepting **Fracastorius M** (6 km) halfway along. Some 20 km east of Fracastorius M, a small crater chain (13 km long) approaches rille from the south, but since its half-dozen individual crater components are unresolvable through amateur instruments, it has the appearance of a branch of the main rille. A number of unusual-looking conjoined craters lie around Fracastorius's walls. In the west, with the flooded, misshapen craters **Fracastorius E** (9 km), a group of three, small, flooded craters are joined, and to its south lies **Fracastorius H** (21 km), an overlapping, flooded crater trio. Indenting Fracastorius's western rim, **Fracastorius D** (29 km) is the largest of a linked chain of craters (75 km long) that runs south through **Fracastorius Y** (12 km). On the opposite side of Fracastorius, adjoining the northern tip of the main crater wall, lies an unnamed flooded crater (14 km) shaped like a cloverleaf. Further north, **Rosse** (12 km), a deep bowl crater, lies along a wrinkle ridge. Under a high Sun, one of Tycho's rays can be traced across the western wall of Fracastorius, passing 10 km west of Rosse. Another ray – perhaps also originating from Tycho – begins a short distance northeast of Rosse, crossing to the northeastern edge of the mare and gradually broadening along its path.

East of Fracastorius, Mare Nectaris's coastline is poc kmarked with a dozen small (sub-25 km) sharp-rimmed craters. To their north, a collection of low wrinkle ridges runs north across the mare to **Bohnenberger** (33 km), a crater with a hilly floor that is cut through by several small rilles, the largest of which (14 km long) is visible through a 150 mm telescope. At 130 km to the south, **Santbech** (64 km), a deep crater with an offset central peak, occupies a smooth, gray plain that links southeastern Mare Nectaris to southwestern Mare Fecunditatis.

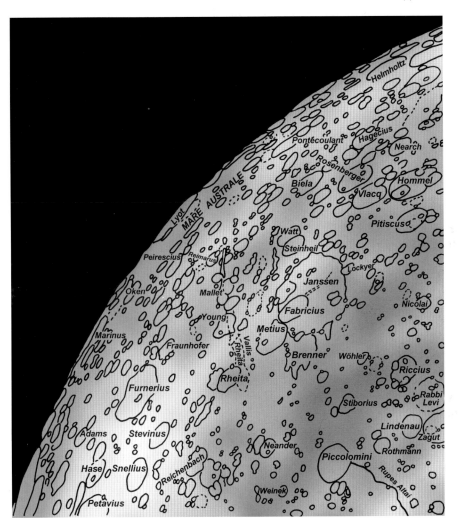

Area Sixteen

Mare Australe (Southern Sea), a collection of large, dark-floored, flooded craters spanning a vast circular region, occupies much of the libration zone along the far-southeastern limb of the Moon. Between the large craters **Janssen** and **Furnerius** lies **Vallis Rheita**, an impressive valley made up of a chain of large craters. **Stevinus A** and **Furnerius A**, both relatively small craters, are the sources of the area's two brightest ray systems. Sunrise takes place between full Moon and around 4 days. Sunset is from full Moon to around 19 days.

A large area south of Piccolomini (see Area Thirteen) is rough and striated with broken lines of ridges and hills likely to have been sculpted during the formation of the Nectaris impact basin. **Brenner** (97 km) is highly eroded and lies at the southern end of a prominent wiry ridge that runs from the southeast of Piccolomini, its shadow visible under a low morning illumination. **Metius** (88 km), to its immediate east, is prominent, sharp-rimmed and deep, with a group of rounded central hills and **Metius B** (17 km) on its northeastern floor. **Fabricius** (78 km) adjoins Metius's southwestern wall. Fabricius itself, and its battered eastern neighbor **Fabricius A** (55 km), lie entirely within the confines of the large crater **Janssen** (190 km). Two large, parallel mountain spines run across Fabricius's floor; the central spine is 11 km long, and the northern one runs for 29 km across the floor to the inner northeastern wall. Fabricius' rim is slightly irregular in outline, and a breach in its south wall extends into a lobe that pushes across Janssen's hilly northeastern floor.

Janssen is one of the largest features in this region of the Moon, and is easily visible through binoculars when it is illuminated by a low morning or evening Sun. It has a distinctly hexagonal outline, and although its rim does not rise particularly high above the surrounding landscape, its outline is enhanced by the presence of Fabricius and Fabricius A on its floor, indenting the northeastern rim, a broad inner eastern and southern wall and a sharp western rim. Janssen's floor is rough and hilly, particularly in the north, where it is bisected by a broad (9 km wide) eroded valley, which appears to be a secondary impact structure, radial to the Nectaris basin. The largest linear rille in the **Rimae Janssen** system extends west from the hilly northern part of the floor, across the older valley and over to Janssen's inner western wall. Another rille extends from the base of the inner southern wall, curving slightly along its course, to meet the main rille. These features are easily detected using a 150mm telescope.

The prominent conjoined duo **Steinheil** (67 km) and **Watt** (66 km) lie a short distance southeast of Janssen. Steinheil has a smooth, dark floor, and it slightly overlaps the northern wall of Watt, a more eroded crater with a ridged floor. Some 100 km further to the south, **Biela** (76 km) is a prominent crater with internally terraced walls and a slightly off-center group of mountains on its flat floor. **Biela C** (28 km), superimposed upon Biela's northeastern rim, is noteworthy because of its floor, piled high with masses of rubble.

Vallis Rheita (500 km long), a prominent chain of craters gouged deeply into the highlands east of Janssen, is the longest valley on the Moon's near side. It runs from the western flanks of the prominent crater **Rheita** (70 km), slicing through the eroded craters **Young** (72 km) and **Mallet** (58 km) to taper off in a narrower valley west of **Reimarus** (48 km). Like many of the other linear features in this area of the Moon, it is radial to the Humorum multiring impact basin and is probably a secondary impact feature.

Furnerius (125 km) is the southernmost of a prominent line of large craters – Langrenus, Vendelinus, Petavius and Furnerius – that run along the terminator of the 3-day-old Moon. Its rim, inner walls and much of its floor have a sandblasted texture, but a narrow. central strip of the floor from the inner southern wall to the east of **Furnerius B** (21 km) is dark and smooth, punctured at its very center by a small, solitary mountain peak. **Rima Furnerius** (70 km long), a linear rille visible through a 150mm telescope, cuts northwest–southeast across the crater's northern floor. Another narrow linear rille, **Rima Hase** (275 km long) cuts across the uplands east of Furnerius, from Hase D (see Area Fifteen) to **Marinus C** (35 km), just on the edge of the southeastern libration zone.

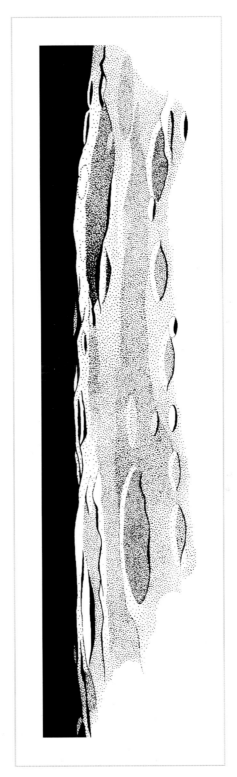

Fig. 7.71. The region of northern Mare Australe, from craters Oken to Lyot. Stippled observational drawing. Credit: Grahame Wheatley

Further east lies **Mare Australe**, the name given to a vast circular area some 500 km in diameter that is composed of dozens of large, flooded craters. Its total surface area measures more than 150,000 sq km. Mare Australe straddles the 90°E line of longitude, and it occupies the libration zone; most of it, with the exception of its far-eastern reaches, may be seen during a favorable libration. An unfavorable libration will render Mare Australe completely invisible beyond the southeastern limb. Notable dark-floored craters within Mare Australe on the mean lunar near side include **Oken** (72 km) on its northwestern border and **Lyot** (141 km) to its south. Beyond 90°E, and visible only during a favorable libration, is the prominent central peak crater **Jenner** (71 km) and the large, flooded crater **Lamb** (104 km) to its east. Beyond Mare Australe's southeastern border, **Vallis Schrödinger** (310 km long) lies between **Moulton** (49 km) and the northeastern ramparts of the vast basin **Schrödinger** (310 km), bisecting **Sikorsky** (98 km) in-between. Vallis Schrödinger is the largest of a number of large, linear valleys radial to Schrodinger, each a secondary impact feature made up of a chain of craters. Vallis Schrödinger is one of the most elusive features on the Moon, requiring a very favorable libration and evening illumination (just after full Moon) to discern.

Chapter 8

Advanced Lunar Research

Transient Lunar Phenomena (TLP)

From time to time, experienced lunar observers have reported transient changes on the Moon. These changes are known as TLP (transient lunar phenomena) and reportedly take the form of localised flashes, glows, obscurations, darkenings and changes of color that take place on a temporary and rather unpredictable basis. Some colored glows appear to be associated with specific features, notably in areas that have undergone volcanic activity, or features that display faults or along faulted mare borders.

A number of astronomical societies around the world have implemented their own TLP watch programs and TLP alert networks. Well-planned though these projects are, lunar coverage is by no means as complete or as continuous as it could be, so there is ample opportunity for the lunar observer to make a significant contribution to this field of research.

Reports are allocated a weighting dependent upon the following factors:

1. *Observer's experience.* An uncorroborated TLP report submitted by an inexperienced lunar observer using a small telescope is not likely to be afforded the same weighting as an observation of the same phenomenon made by an experienced lunar observer who has logged many hours at the eyepiece in a painstaking TLP search.

2. *Corroboration.* Reports of identical phenomena submitted independently by two observers at different locations. Independent confirmation of what appears to be the same anomalous lunar event is perhaps the most important criteria, alongside photographic evidence, in the weighting system.

3. *Telescopic aperture.* An important factor in allocating a weighting to observations. For example, a single unconfirmed report of a vague red glow near Aristarchus, seen through a 50mm refractor, will be given less weighting than an observation of the same kind of thing made with a 250mm reflector.

4. *Special equipment employed.* Colored filters can enhance the visibility of anomalous colored areas on the lunar surface. Photographs, CCD images and videos of apparently anomalous lunar events are exceptionally rare. A permanent record can be analyzed in great detail to discover the TLP's exact location, its shape and extent, brightness and possible coloration.

5. *Conditions of observation*. The Moon's height above the horizon, atmospheric visibility, cloud cover and transparency all contribute to how clearly the Moon's surface appears to an observer.

6. *The type of phenomena reported and its duration*. To use an extreme example, a report of a vague colored glow near Aristarchus will tend to be taken more seriously than a report of a vast silver saucer-shaped object that temporarily materialized over Mare Imbrium.

Features with Reported TLP Associations

Agrippa
Archimedes
Aristarchus
Atlas
Alphonsus
Bullialdus
Censorinus
Copernicus
Eratosthenes
Gassendi
Grimaldi
Herodotus
Kepler
Linné
Manilius
Mare Crisium
Menelaus
Mons Piton
Mons Pico
Picard
Plato
Posidonius
Proclus
Promontorium Laplace
Schickard
Theophilus
Tycho
Vallis Schröteri

Of these features, Aristarchus has by far the most TLP reports – more, in fact, than all the other features put together. The TLP reported in and around Aristarchus have been chiefly brightening events, with a number of colored anomalies and obscurations of surface detail.

Blinking Moon

A number of observers employ colored filters to determine anomalous red- or blue-colored transient events on the Moon. By quickly alternating a red and blue filter in the telescopic light path, faintly colored areas on the Moon can stand out more by appearing to flicker. A red area will appear brighter when viewed through a red filter, and darker when seen through the blue filter. It is recommended that two specific colored filters are used for the best results – a Wratten 25 red and a Wratten 38a blue filter. It is entirely possible to mount these filters side-by-side on a card, and manually alternate them in front of an eyepiece that has a good level of eye relief. Such a manual technique may be just as effective as using a purpose-built filter wheel, but will require some practice and coordination. Atmospheric conditions in the observer's locale are capable of causing a blink effect on features across large parts of the lunar disc, an area displaying true coloration will appear to blink while features of a similar albedo in the nearby region do not. There are a number of features on the Moon that will appear to blink naturally, among them being the southwestern part of the crater Fracastorius and a section of the western wall of Plato.

A special filter wheel called a "crater extinction device" or a calibrated variable-density filter will enable the brightness of an individual lunar feature to be measured according to the point at which it ceases to be visible. With practice, a feature's apparent brightness during the lunar day can be ascertained, along with features in the target's immediate vicinity. Using the device, it is possible to determine whether a feature is exhibiting a genuine brightness anomaly by comparing it with the brightness of other features in the immediate area.

Banded Lunar Craters

A large number of lunar craters display dusky albedo banding in their inner walls, most clearly visible when the crater is devoid of shadow. Banding is not necessarily associated with any obvious topographic features, and it may be entirely independent of any terracing in the crater's internal walls. There are a number of possible causes of banding. In some instances, banding may be produced by a loose covering of darker regolith produced by a small landslide down the inner slopes of a crater. Certainly, a number of bands appear to be associated with marked irregularities in the crater's rim. Some banding features may be lava flows that have flowed over the crater's rim and into the crater. There is evidence for this in a number of craters surrounded by lava flows in the maria, whose walls have been visibly breached in places. Alternatively, banding may reflect actual variations in the composition of the lunar crust. For example, an intrusive magmatic dike that has been exposed by the crater's excavation may appear as a distinct, dark line. In some cases, dusky banding may be caused by the contrast between the actual dark color of the crust and overlying ejecta material of higher albedo. This ejecta may originate from the parent crater, or from an impact feature further afield.

The most notable example of a banded crater is Aristarchus, on whose inner western wall banding may clearly be discerned through telescope whenever the inner western slopes are illuminated. Many other craters display similar banding features. According to Abineri and Lenham (*BAA Journal*, March 1955), banded craters have five basic appearances.

Type 1: Aristarchus

Aristarchus is the largest example of this type of banded crater. The others are similarly very bright, rather small, with small, dark floors surrounded by broad, bright walls. The bands generally appear to radiate from near the centers of the craters. Bright ray systems or ejecta collars surround a number of these craters. This is the most widespread type of banded crater. Examples include the following:

Airy B
Albategnius F
Aristarchus
Aristillus
Beaumont R
Biot W
Bode ph
Boguslawsky C
Bohnenberger G
Cayley
Chevalier A
Cichus B
Fourier E
Fracastorius N
G. Bond
Gutenberg A
Henry Frères B
Isidorus D
Isidorus E
Janssen
Kies A
Luther
Mersenius S
Rabbi Levi F
Rosse
Riccius E
Silberschlag W
Sacrobosco C
Schomberger A
Stiborius G
Strabo
Theon Senior
Theon Junior
Turner E
Wilhelm G
Wilkins A

Type 2: Conon

Rather dull craters with large, dark floors and narrow walls. Very short bands can be discerned on the inner walls, but they are unable to be discerned on the floors. Despite their shortness, the bands appear radial to the crater's center. Examples of Conon type banded craters include:

Adams B
Boussingault D

Conon
Cuvier E
Cuvier B
Cyrillus R
Fracastorius B
Fracastorius K
Guericke B
Haidinger
Henry Frères
Kepler
Kircher W
König
Marinus A
Menelaus
Nöggerath J
Pytheas
Reichenbach C
Theaetetus
Timocharis
Wilhelm A
Wilhelm B

Type 3: Messier

A broad dark band crosses the crater floors. A rare type of banded crater, examples include:

Cavendish E
Messier
Messier A
Wilhelm K

Type 4: Birt

Long and often curved dark bands appear to radiate from a dusky off-center area. The parent craters' brightness and size are similar to those of type 1. Examples of Birt-type banded craters include:

Abenezra B
Birt
Chladni
Cichus C
Darney
Davy A
Glaisher W
Hercules D
Marco Polo
Maury
Nicollet W

Opelt E
Ptolemaeus A
Thebit A

Type 5: Hippalus A

This type of crater has bands that appear to radiate from near the wall inside a dark section of the crater floor, and are visible on the dull and bright parts of the floor. The rarest type of banded crater, only two have been identified:

Berzelius H
Hippalus A

The Measure of the Moon: Calculating Crater Depths and Mountain Heights

Once the pursuit of many a dedicated professional astronomer armed with very large observatory telescopes and bifilar micrometers, measuring the depths of lunar craters and the heights of lunar mountains can be an enjoyable and rewarding pursuit for today's amateur lunar observer. Since the height of the Sun above any part of the Moon at any time is already known with considerable accuracy, the method uses the length of a shadow cast by a lunar feature to determine its vertical height above the terrain upon which the shadow is cast.

Computer software, such as Harry Jamieson's *Lunar Observer's Toolkit*, enables the observer to bypass much of the involved mathematics that were once necessary to calculate vertical heights on the Moon. All that is required is to input several key pieces of information, including the observer's exact geographical location and altitude, the date and UT of observation, the exact selenographic coordinates of the feature casting the shadow and the measured length of the shadow.

Shadow measurements can be derived through visual observation or by analysing CCD images. Using simple visual estimates of a shadow's length in relation to nearby features, a reasonable estimate of the feature's height can be made. More accurate visual observations can be achieved using the drift method, where the shadow being measured is allowed to drift across the cross hair of a high-power eyepiece and its passage timed. A nearby feature of known size is also allowed to drift across the eyepiece and timed, and by comparing the two timings it is possible to derive a good figure for the length of the shadow. Using an eyepiece with a fine, graduated reticle, an even more accurate determination of the length of a lunar shadow can be made.

Calculating Feature Heights

In order to accurately calculate the heights of features mathematically, it is necessary to know both the selenographic coordinates of the feature observed and the Sun's selenographic colongitude to determine the angle of the incident sunlight above the feature. Following is a simplified method of calculating a feature's height based on a measurement of its shadow:

Make an observation of the shadow cast by a hypothetical mountain peak located at 10°N, 10°W. Using the drift method, its shadow has been estimated to be 30 km long on 14 February 2004 at 01:00 UT.

The Sun's selenographic colongitude at 01:00 UT = 187.4°.

The selenographic longitude of the evening terminator is calculated to lie at 7.4°W.

At the location of the feature, 10°W, the Sun is 2.6° above the horizon.

To find the height of the feature, multiply the tangent of the Sun's angular height by the shadow length:

Tan 2.6° = 0.041   (1)

0.041 × shadow length = 0.041 × 30.75 km = 1.26 km   (2)

Height of mountain = 1,260 m   (3)

This calculation above is, of course, the most simple means of gauging a feature's height by trigonometrical means.

Another method of determining the height of features involves imaging the Moon and determining a shadow's length by its measured size on the image. This requires the exact scale of the image to be known and usually involves incorporating the Moon's apparent angular diameter at the time of the image into the equation. Importantly, a correction factor needs to be introduced into the equation to account for the curvature of the Moon's surface since the further a feature is in longitude from the mean center of the Moon's disc, the shorter a shadow of given length will appear to be. Good quality digital images are best to work from since the shadows are usually more clearly defined than in conventional photographs owing to the shorter exposure time required and less obvious effects of image drift on the clarity of the image. Determining feature heights is a useful educational process, but other than this the exercise has little scientific use today.

Chapter 9

The Lunar Observer's Equipment

An Eye for Detail

Regardless of whether a lunar observer owns a tiny reflector on a basic altazimuth mounting or a large, expensive apochromatic refractor on a sturdy computer-controlled mount, by far the most important optical equipment belonging to any lunar observer is a small but powerful pair of binoculars, the eyes. Looking after these precious little instruments carefully will allow the owner to enjoy the beauty of the Moon's surface throughout his or her lifetime.

Although the light of the Moon can appear intensely bright through the eyepiece, moonlight – even the vast quantities of lunar photons gathered and focused by a really big telescope – cannot cause any damage to one's eyesight. Moon filters that are inserted into an eyepiece are only intended to reduce glare and enhance contrast, they're not meant to prevent eye damage in the way that solar filters are.

The human eye is a spherical organ measuring around 4 cm across. A transparent membrane at the front of the eye, the cornea; focuses light rays through the aqueous humor, a chamber filled with transparent fluid, through an opening in the iris, the pupil, and on through a clear crystalline lens behind it. This lens focuses light across a chamber filled with vitreous humor, a clear jelly that gives the eyeball its rigidity, and an upside-down image is projected onto the retina inside the back part of the eyeball. The retina contains millions of light-sensitive cells, rods and cones, which convert the light into electrical impulses sent directly to the image-processing center of the brain through more than 1 million nerve cells in the optic nerve. The brain automatically flips the image the right-way-up and processes it.

At the very center of the retina lies the fovea, the spot with the greatest concentration of cones, allowing high-detail color vision at the center of the field of vision. There are three types of cone cells – these are sensitive to red, yellow-green and blue light – but the cones are not triggered in low light-level situations. This does not affect the lunar observer, since the light levels in a telescopic image of the Moon are always high enough to allow detailed color vision. However, in low-light circumstances, for example, when attempting to view dim galaxies, only the rod cells lying around the fovea, away from the center of the field of view, are triggered. To get the best view of a faint, deep-sky object, the observer must look about 15° away from its actual location, a technique known as "averted vision." The rods do not deliver such detailed images, nor are they capable of distinguishing between different colors, so most deep-sky objects appear as black and white when viewed visually.

One aspect of the eye's structure that does affect lunar observing to some degree is the presence of a blind spot in each eye, caused by the lack of photoreceptive cells in the part of the retina where the optic nerve intrudes. Blind spots lie both on the left side of the left eye's field of view and the right side of the right eye's field of view; their presence can sometimes be revealed when viewing lunar features through a fairly wide-angle eyepiece. For example, a small crater lying some distance from a feature under scrutiny can appear to completely vanish from sight, while features in the terrain surrounding the crater remain visible; of course, this effect is noticeable only through averted vision, and is cancelled out immediately when the eye flicks onto the affected area. The following experiment demonstrates the complete blindness of the blind spot, and it usually astounds anyone first trying it: Cover your left eye with your left hand and, with your right eye, slowly scan the area a few degrees away from the left of the Moon. You will eventually be able to find a spot where the Moon has completely vanished from sight – all that can be seen is a glow surrounding a blank spot in the sky. The actual area covered by the blind spot measures about 6° across, 10 times the apparent diameter of the Moon.

Many people suffer from floaters – minute dark flecks, translucent cobwebs or clouds of various shapes and sizes, which become visible when viewing a very bright object like the Moon. Floaters are the shadows cast onto the retina by the remnants of dead cells floating in the vitreous humor. Everyone has floaters, and they can be annoying because they can obscure small lunar features. Most people put up with them. Floaters increase in number with age, and laser eye surgery or a vitrectomy can remove them if the condition causes severe disruption of normal vision.

Annual eye checkups are recommended, as some unsuspected but treatable medical conditions may come to light as a result. Smoking is detrimental to astronomical observation, let alone the smoker's general health. Eyes use nearly all of the oxygen that reaches them through the blood vessels, and smoking reduces the blood oxygen content by around 10% of a pack a day – the carbon monoxide in cigarette smoke attaches to the hemoglobin in red blood cells more readily than oxygen itself. Drinking alcohol while observing decreases the detail seen on the Moon in direct proportion to the amount of alcohol consumed, and the observer's efficiency is totally compromised when, despite one's best efforts, the eye cannot be kept close to the eyepiece. Alcohol dilates the blood vessels, and though it may make the consumer feel warm for a while, the extra heat loss from the body can be dangerous on cold nights. So, alcohol ought to be avoided until after the observing session when one is safely indoors. Foods beneficial to healthy vision include a variety of fruits and vegetables, especially dark-colored ones like carrots and broccoli, good sources of beta-carotene and many carotenoids. These substances aid night vision and help maintain good vision. Vitamin C helps protect the eyes against ultraviolet radiation and, being an antioxidant, it inhibits the natural oxidization of cells. The development of cataracts and age-related macular degeneration (AMD) is inhibited by Vitamin E, a substance found in wheat germ oil, sunflower seed and oil, hazelnuts, almonds, wheat germ, fortified cereals and peanut butter. Finally, zinc helps maintain a healthy retina and can play a part in preventing AMD. Zinc is found in wheat germ, sunflower seeds, almonds, tofu, brown rice, milk, beef and chicken. Visual acuity actually diminishes if blood sugar levels are low, so a small snack during the observing session is both pleasant and beneficial.

Binoculars

Binoculars are somewhat underrated by amateur astronomers, and few would seriously consider using them to regularly observe the Moon. Yet binoculars have many advantages over telescopes. They are usually far less expensive, are easier to carry around (even with a stand), deliver a wide field of vision and are more robust than telescopes, capable of withstanding occasional knocks.

Opera glasses (also known as Galilean binoculars) are the simplest binocular design. In their most basic form, they consist of small biconvex objective lenses and biconcave eye lenses, and these form an upright image. Opera glasses have a short focal length, and produce a low magnification and a narrow field of view. Their unsophisticated optical configuration produces false color (chromatic aberration), which becomes obvious when bright objects like the Moon are viewed. Often made to fold into a compact size when not in use, opera glasses are lightweight and easy to carry around in a coat pocket or handbag. Opera glasses will reveal the Moon to be a rugged globe pitted with craters and swathed in large, dark lava plains, some of which are surrounded by large mountain ranges. Lunar observers will find them useful to keep handy, since they can be used for quick peeks at the Moon and to discern which features might be on view along the lunar terminator, enabling detailed telescopic observing sessions to be planned in advance.

To discern fine lunar detail it is important to hold binoculars as steady as possible, either propped up firmly against a solid object (like a car roof) or, better still, fixed to a tripod or a dedicated binocular mount. A steady field of view adds enormously to the observer's pleasure; those restricted to hand-held viewing may view the skies for a few minutes at a time, but observers who are able to hold their binoculars steady are likely to observe for far longer periods of time. When the Moon is steadily positioned in the field of view, a quality pair of binoculars is capable of revealing a tremendous amount of lunar detail.

The power of binoculars is identified by two figures that denote their magnification and the size of their objective lenses: 7×30 binoculars give a magnification of 7× and have 30mm objective lenses. Small- to medium-sized binoculars (with objective lenses from 25–50mm in diameter) usually deliver low magnifications that range between powers of 7 and 15×. Since the Moon is such a bright object, and binoculars have such low magnification, the difference in the amount of lunar detail visible through, say, 7×30 binoculars and 7×50s may not be immediately noticeable.

The true field of vision (the actual area of sky) that is observed through binoculars decreases with magnification. My own 7×50 binoculars (a budget brand, but of good optical quality) deliver 7× magnification and a true field of view some 7° wide, so the Moon's half-degree diameter takes up one-fourteenth of the field diameter. The apparent diameter of the Moon through 7× binoculars measures about 3.5°, and through 15× binoculars 7.5°. My 15×70 binoculars have a true field of view of 4.4°, some nine times the diameter of the Moon. Binoculars equipped with wide-angle eyepieces (usually high-end instruments) produce larger actual fields of vision; apparently greater than 60°. With their wide true fields of vision, binoculars take in a large area of the sky around the Moon. Often, the glare of the Moon drowns our all but the brightest stars in its vicinity, but stunningly beautiful views can be had when the Moon is at its crescent phase, its unilluminated side glowing blue with earthshine and in the vicinity of bright-star groupings and the occasional planet. Binoculars can be used to observe the brighter lunar occultations of stars and planets. I would recommend that lunar eclipses be viewed through binoculars rather than through a single telescope eyepiece; using both eyes adds to the observer's enjoyment, the subtleties of the colors in the umbral shadow stand out more, and the sight of the eclipsed Moon set against the stars can appear almost 3-dimensional when viewed with two eyes.

Small binoculars will reveal all the near-side maria (along with some marial detail, such as wrinkle ridges, when near the terminator), all the Moon's major mountain ranges and several hundred craters. The effects of the Moon's libration are immediately noticeable, and it is possible to identify Mare Humboldtianum near the northeastern limb and Mare Orientale near the southwestern limb during favorable librations. Binoculars that are capable of zooming from low to high power enable closer scrutiny of the lunar surface. Large binoculars (those with 60mm objectives and larger) with high-magnifications provide detailed views of the lunar surface. A pair of giant 25×100 binoculars will deliver wonderful views of the Moon, even though considerable internal reflections might be visible in budget models. Any binoculars with more than $10\times$ magnification must be supported firmly, however lightweight they may be, because higher magnifications exacerbate any slight movement of the observer's body. There is an exception – image-stabilized binoculars. First introduced in the early 1990s, image-stabilized binoculars eliminate the minor shakes of the user's body. At first glance, they resemble regular binoculars of the same aperture but weigh slightly more, and they can be used like normal binoculars. At the push of a button they deliver sharp, vibration-free views courtesy of moving optical elements (most of them require batteries). Stabilized binoculars typically offer quite high-magnifications (up to $18\times$), with apertures from 30 to 50mm – nice for viewing the Moon.

Some viewers of the Moon are tempted to seek the best of both worlds by having the convenience of binoculars with a zoom facility. A typical pair of zoom binoculars can be adjusted to magnify from, say, $15\times$ to as much as $100\times$. On the face of it, such an instrument would be ideal for general lunar observation, but there are drawbacks. When set at a low magnification, the apparent field of view in zoom binoculars is usually minuscule – perhaps as small as 40°, far smaller than the apparent field visible in a pair of regular binoculars of comparable magnification. Zooming involves physically altering the distance between the lenses inside the eyepiece using an external lever. This is not such a simple operation, as some re-focusing is usually necessary after a change in magnification. Crucially, the alignment of the left and right optical systems in any zoom binoculars needs to be absolutely exact in order that the brain can produce a merged image from two separate high-power images; this is where most budget zoom binoculars prove wanting. Even if a pair of binoculars does provide good high-magnification views of the Moon, the rotation of the Earth makes the Moon appear to move across the field of view at a rate of its own diameter every 2 minutes or so. For example, at a magnification of $50\times$, an apparent field of 50° will equate to a true field of view of 1°, and the Moon will move from one edge of the field to the other in just 2 minutes. This means frequent adjustments to the mount if the Moon is to be scrutinized for any length of time.

Numerous optical configurations are used in binoculars, and this is obvious when one looks at the wide variety of different shapes and sizes available. As is usually the case, you get what you pay for, but providing they are purchased from a reputable optical dealer, the quality of budget optical instruments these days is usually quite good. The quality of the optical system, the optical materials used and the build of binoculars is, however, noticeable when budget binoculars are compared with high-end binoculars. High-end binoculars use the best optical glass in their object glasses, internal prisms and eyepiece lenses, and they are figured and aligned to exacting standards. The optical surfaces are usually multi-coated to minimize reflections, and internal baffles block stray light and internal reflections, providing better contrast.

Most binoculars use glass prisms that fold the light between the objective lens and eyepiece lens, and they produce a right-way-up image, which is, of course, essential for everyday terrestrial use. The 7 × 50 binoculars are ideal general-purpose astronomy binoculars. They deliver a wide field of view and have a magnification low enough for the observer to peruse the skies for short periods without requiring the use of a binocular support. These binoculars have an exit pupil of 7 mm. The exit pupil is the diameter of the circle of light that projects from an eyepiece into the eye, and its size can be derived from the aperture of the binoculars divided by their magnification. As the dark-adapted eye has an average size of 7 mm, the optimum size of exit pupil for deep, dark-sky astronomical use measures 7 mm. Because the Moon is such a bright object, and the observer's pupil contracts to a smaller size when it is viewed, this is not relevant to most lunar observation.

There are two basic types of prism binoculars – Porro and Roof. Until a couple of decades ago, most binoculars were of the Porro prism type. Most often, these binoculars have a distinct W shape, produced by the arrangement of the prisms to fold the light from the widely separated objective lenses to the eyepieces. American-style Porro prism binoculars are sturdy designed, with prisms mounted on a shelf inside a single molded case. German-style Porro prism binoculars feature objective housings that screw into the main body containing the prisms; their modular nature makes them more susceptible to going out of collimation after knocks. In recent years, a new style of small binoculars has appeared; it uses an inverted Porro prism design, leading to a U-shaped shell with objective lenses that may be closer together than the eye lenses. The majority of small binoculars produced today are of the Roof prism design, popular because they are compact and lightweight. Roof prism binoculars often have a distinctive H-shape that looks like two small telescopes placed side-by-side. A casual observer might think this indicates a straight-through optical configuration without any intervening prisms. Because of the way that Roof prism binoculars fold the light, they offer generally less contrasty views than Porro prism binoculars.

Telescopes

Anyone who has marveled at the sight of the Moon through binoculars will have an overwhelming desire to view the lunar surface at high-magnification through an astronomical telescope. There is nothing to compare with a close-up view of the Moon through a telescope on an evening of good seeing conditions: when seen up close, the mountains, valleys, craters and lava plains of the Moon appear more 'real'

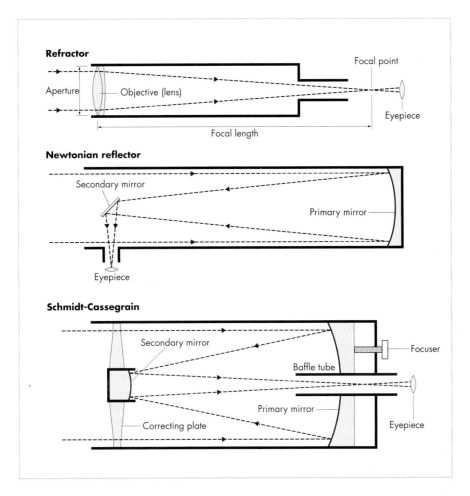

Fig. 9.1. Simplified cross sections showing the light path through the three major types of telescope used by amateur astronomers – a refractor, a Newtonian reflector and a Schmidt-Cassegrain. Refractors, with their fixed optics and enclosed system, are the least demanding of maintenance. Newtonian reflectors require occasional optical collimation, cleaning and realuminising of their mirrors. Schmidt-Cassegrain telescopes have an enclosed optical system, but their secondary mirrors sometimes require collimation to keep the light path perfectly aligned.

than any other telescopic sight in the heavens. Every amateur astronomer who has owned a telescope will have turned it towards the Moon, and few have ever been disappointed at the view. There is no substitute for observing the Moon directly through a telescope eyepiece. If asked to choose between enjoying the Moon visually or in the comfort of a warm room viewing a live image from a CCD camera, the majority of amateur lunar observers would undoubtedly choose to be out in the field with the telescope, allowing their retinas to be bombarded by photons that have arrived directly from the lunar surface. Many public star parties arranged by local astronomical societies are purposely timed to take place when there is a Moon to be seen in the evening skies, proof that the Moon really is the most spectacular sight in the skies. People may strain to see cloud belts on Jupiter, or the dusky markings on Mars, but the Moon is packed with easily observable detail, especially along the terminator where most shadow is produced by topographic features.

One of the main reasons that people buy their first telescope is to see detail on the Moon – indeed, advertisements for many entry-level instruments play on this fact by including stunning images of the lunar surface. However, choosing the right telescope can be a difficult task for a novice. What may be a good telescope for low-power, deep-sky observation may not be the best one for observing the Moon and planets. An amazing array of telescopes of various kinds are advertised in astronomy magazines. Thankfully, these days the optical and build quality of most of these telescopes (an ever increasing proportion Chinese imports) is quite acceptable for general astronomical observation and viewing the Moon at low to medium powers.

Regardless of a telescope's aperture or physical size, the most important thing to be aware of when purchasing any telescope is its optical quality. It is not advisable to buy a new telescope from any source other than a reputable telescope dealer, and I recommend that sensational newspaper advertisements, department stores and general electrical stores ought to be avoided. Unfortunately, newspaper advertisements announcing massive stock clearances of telescopes and binoculars are often full of outrageous hyperbole, with claims of extravagant magnifications and the ability of the instrument to show all the wonders of the Universe. These ads attempt to blind the novice with 'science' in an attempt to hide the fact that their instruments may be made entirely of plastic – even the lenses! Such optical monstrosities are useless for any kind of observing, and the poor view that they deliver can be enough to put the novice off astronomy altogether! Telescopes sold through large shopping mall home goods retailers and department stores are usually overpriced. Moreover, large general retailers are not overly concerned about the optical quality of their goods, and the sales assistant will probably be unable to inform the buyer of the suitability of the telescope for astronomical purposes. The ultimate test of a retailers' confidence in the quality of their goods, and a measure of their customer friendliness, is to ask to examine and make a quick observational test of any telescope that you intend to buy. Any major flaws in the optics, or dents and defects on the exterior of the instrument, are quickly revealed under the bright lights of the store.

Any reputable company that specializes in selling and/or manufacturing optical instruments will give by far the best deal in terms of price, service and advice. All the major astronomical equipment retailers advertise in astronomy magazines, and most of them produce a product catalogue that can be browsed online or in printed form.

Refractors

Asked to imagine a typical amateur astronomer's telescope, the refractor jumps to mind. Refractors have an objective lens and an eyepiece at either end of a closed tube. Light is collected and focused by the objective lens (the light is refracted, hence the name of the optical system), and the eyepiece magnifies the focused image. Reference is often made to a telescope's focal length, the distance between the lens and the focal point, expressed as a multiple of the lens diameter or in millimeters. A 100mm f/10 lens has a focal length of 1000 mm. A 150mm f/8 lens has a focal length of 1,200 mm. Eyepieces also have focal lengths, but these are always expressed in millimeters, and never as a focal ratio, there is no such thing as an f/10 eyepiece, for example.

Galilean telescopes are the simplest form of refractor, with a single objective lens and a single eyepiece lens. They suffer greatly from chromatic aberration caused by the splitting of light into its component colors after it is refracted through glass, and spherical aberration caused by light not being brought to a single focus. Through a Galilean telescope the Moon is surrounded by fringes of vividly colored light, and the whole image appears washed out and blurry. Cheap, small telescopes attempt to alleviate the worst effects of aberration by having large stops placed inside the telescope tube, preventing the outer parts of the cone of light from traveling down to the eyepiece. This crude trick simply makes a poor image appear slightly less poor, and the presence of the stop reduces the aperture of the instrument, as well as light grasp and resolving power.

Small though they may be, good-quality astronomical finderscopes and monoculars should not be confused with Galilean telescopes. Finderscopes are low-power refractors that are attached to, and precisely aligned with, larger telescopes in order that the observer can locate celestial objects. When centered on the finder's cross hairs, the object is also visible at a higher magnification in the main instrument. Finderscopes have achromatic objective lenses (typically 20–50mm) and have fixed eyepieces that can be adjusted for focus. Straight-through finderscopes deliver an inverted view, so they are unsuitable for terrestrial use. Monoculars are little hand-held telescopes with small (typically 20–30mm) achromatic objective lenses. Monoculars use Roof prisms to deliver low power, right-way-up views. They can be carried in a coat pocket and are great for cursory lunar observation. With their large true fields of view, bright stars and planets in the immediate vicinity of the Moon can be seen.

A well-made telescope of any size is capable of providing a really pleasing view of the Moon. Those who claim that any telescope with an aperture smaller than 75mm is useless for lunar observation misunderstand the main reason why most people observe the Moon – for the sheer pleasure of scanning their eyes along the rugged lunar terminator, to be enthralled by the magnificent desolate expanses of the marial plains and to be astounded by the grandeur of our sister planet. Every type of lunar feature, from the vast maria to some of the intricate narrow rilles, can be discerned through a 40mm refractor. Although a small telescope is incapable of revealing some of the finer detail on the Moon, there is enough detail available to keep a novice enthralled, busy learning his or her way around the Moon, for a very long time.

On nights of poor seeing, when the atmosphere is shimmering and the stars scintillate wildly, attempts to observe the Moon at high-magnification through large instruments can prove futile, as it can take on the appearance of a shimmering, boiling mass, hardly worth bothering to look at. On these nights, a small telescope will sometimes deliver an apparently sharper, more stable image than the large telescope, since a small telescope will not resolve as much atmospheric turbulence as a larger one. Small telescopes have a number of other advantages. Being lightweight, they are eminently portable, and they can be carried around an observing site to avoid local obstacles to the sky, such as trees and buildings. A small, relatively inexpensive telescope is expendable, and for this reason the observer may actually be inclined to use it more often than a "precious" high-end telescope: accidental damage to the external structure or the optics of a cheap telescope is not nearly as soul-destroying as bashing an instrument that cost 10 times as much.

The least expensive small telescopes, including those handheld, old-style brass "naval" telescopes that use draw tubes to focus the image, have a fixed eyepiece that delivers constant magnification. Some fixed-eyepiece telescopes are a little more sophisticated, allowing some variation of the magnification delivered by the eyepiece. A telescope that allows the eyepieces to be interchanged, allowing the magnification to be alternated between low and high powers, will be a somewhat more versatile instrument. Two or three eyepieces, and maybe a magnifying eyepiece called a Barlow lens, are often provided; these are usually small, plastic-mounted eyepieces with a 0.965-inch-barrel diameter, and they are probably going to be of a very basic optical design and of poor quality. These eyepieces may deliver poor-quality views with incredibly narrow apparent fields of 30° or even smaller.

Eyepieces are not accessories – they are as vital to the performance of a telescope as its objective lens or mirror. So, if a small instrument is not performing as well as might be expected, don't get rid of it immediately; replace the eyepieces with some better-quality ones purchased from an optical retailer. The most widely available eyepieces have a barrel diameter of 1.25 inches, and these can be mounted to a .965-inch eyepiece tube using an adapter or a hybrid star diagonal (a mirror that flips the image 90°). Plössl eyepieces deliver an apparent field of about 50°, and quality budget versions of this eyepiece design are available. Good-quality eyepieces can transform a small-budget telescope into a good performer on the Moon at low to medium magnifications (see below for more information about eyepieces).

Quality astronomical refractors have an achromatic objective lens that comprises two specially shaped lenses of different types of glass nestled closely together. These lenses attempt to refract all the wavelengths of light to a focus at a single point. Achromatic objectives do not eliminate chromatic aberration altogether, but, generally speaking, the effects are less noticeable in longer-focal-length refractors. Many budget imported achromatic refractors have focal lengths of f/8 to as short as f/5, and although they do display noticeable chromatic distortion, mainly in the form of a violet fringe around the Moon and the brighter planets, they offer good resolution and good contrast. One inexpensive way of reducing the false color is to use a minus-violet contrast-boosting filter that screws into the eyepiece. Another, more expensive way to minimize chromatic distortion is to use a specially designed lens (such as a "Chromacorr") that attaches to the eyepiece, transforming a budget achromatic refractor into a telescope that approaches the performance of a high-end apochromat.

Apochromatic refractors use special glass in their 2- or 3-element objective lenses to bring the light to a tack-sharp focus, delivering images that are virtually free from the effects of chromatic distortion. Views of the Moon through an apochromat are almost completely free of distortion and are of high contrast, rivaling the kind of view to be had through a good-quality long-focal-length Newtonian reflector (see below). Aperture for aperture, an apochromatic refractor will be more than 10 times as expensive as a budget achromatic refractor.

Refractors require little maintenance. Their objective lenses are aligned in the factory and sealed in a cell, and they are ready for immediate use straight out of the box. There is no reason to unscrew an objective lens and remove it from its cell, although, being naturally curious, a great many amateur astronomers have been tempted to do it just to see how the thing is put together. This is not recommended, as the reassembled lens will not perform as well, often noticeably so. Over time, the external surfaces of lenses can accumulate a fair amount of dust and debris, and great care must be taken when cleaning them. Most lenses have a thin layer of antireflection coating, which can be disrupted if the lens is cleaned improperly.

Under no circumstances should a lens be rubbed vigorously with a cloth. Dust particles should be carefully removed with a soft optical brush or an air puffer, and any residual dirt can be gently removed with optical lens wipes, each used once and applied in a single stroke. Condensation on a lens should be allowed to dry naturally, and never be rubbed off.

Reflectors

Reflecting telescopes collect light with a specially shaped concave primary mirror and reflect it at a sharp focus. Reflected light is free from the effects of chromatic aberration, but they are prone to spherical aberration, more so in shorter-focal-length systems. The most popular reflecting telescope design, the Newtonian, uses a concave primary mirror held in a cell at the bottom of the tube and a smaller flat secondary mirror mounted on a "spider" near the top of the tube, which reflects light sideways through the tube into the eyepiece. The observer does not appear to be looking directly "through" the telescope, but into its side, something that seems to confuse many laypersons! A well-collimated long-focal-length (f/10 and longer) Newtonian will deliver superbly detailed views of the Moon at high-magnification.

Cassegrain reflectors have a primary mirror with a central hole. The primary reflects light onto a small convex secondary mirror, which reflects the light back down the tube through the hole in the primary mirror and into the eyepiece. Prone to the optical aberrations of astigmatism and field curvature, most Cassegrains are large, observatory-sized instruments with focal lengths ranging from f/15 to f/25 – excellent for lunar studies at high-magnification.

Reflectors require much more care and attention than refractors. Vibration or a slight, sudden knock to the tube can cause misalignment of the primary mirror in its cell or the secondary in its spider. Misaligned optics will produce poor quality images, including dimming, blurring and multiple images near the point of focus. A brand new Newtonian is likely to require recollimation in order to align the optical components as precisely as possible. The alignment of both the primary mirrors can usually be adjusted by hand, using three wingnuts at the base of the mirror cell, but the secondary will usually require a small screwdriver or allen key. Collimation can be time-consuming and a little tricky for novices, but new telescopes should be provided with adequate instructions, and there are many internet resources that explain the process in detail. There are products that can achieve good collimation, including laser collimators and a device called a Cheshire eyepiece. Cassegrains are more difficult to collimate than Newtonians. Most reflectors are not sealed when in use, and coated with a wafer-thin layer of reflective aluminum, the primary and secondary mirrors gradually deteriorate by being exposed to the open air. Special coatings can extend the life of a mirror by a factor of 2 or 3. However, all mirrors accumulate a layer of dust and bits of debris over time, and a primary mirror can look disconcertingly filthy when illuminated by a flashlight at night. Debris on the mirror will scatter light, and as the mirror gets grubbier it will be less effective, producing a decline in image contrast. Cleaning the surface of an aluminized mirror must be performed very carefully, since hard bits of debris scraped across the thinly aluminized surface will leave tracks like those of skates on ice. Loose debris can be blown away with a puffer or a canister of compressed air, and the mirror can be cleaned with absorbent cotton and lens-cleaning fluid or lens wipes; this must be performed very gently, with a single stroke per cleaning wipe.

One way of extending the life of a reflector is to stretch a piece of optically transparent film over the telescope aperture in order to seal the top of the tube (the bottom of a Newtonian is often open, allowing the free circulation of air for a better-image quality). This material is available in large sheets that can be cut to fit. For the best image quality, the sheet should ideally be taut, without wrinkles. When the film itself gets dirty, another disc can easily be produced. A well-protected, well-cared-for Newtonian mirror can last more than a decade before requiring realuminization.

Catadioptrics

Catadioptric telescopes use a combination of mirrors and lenses to collect and focus light. There are two popular forms of catadioptric – the Schmidt-Cassegrain telescope (SCT) and the Maksutov-Cassegrain telescope (MCT). SCTs are becoming increasingly popular. Light enters the top of the telescope tube through a large corrector plate, a sheet of flat-looking glass with a large secondary mirror mounted at its center. The corrector plate is actually aspherical in shape, figured to refract the light onto an internal primary mirror, which reflects light onto the convex secondary mirror, which in turn reflects light back down the tube and through a central hole in the primary mirror into the eyepiece. The relatively large size of the secondary mirror in a SCT produces a degree of diffraction that can slightly affect image contrast. An optically good, well-collimated SCT will deliver superb views of the Moon. Moreover, due to their design, a number of useful accessories can be attached to the "visual back" (the part of the telescope that the eyepiece normally fits into). Those useful to the lunar observer include filter wheels, SLR cameras, digicams, camcorders, webcams and CCD cameras.

MCTs use a spherical primary mirror and a deeply curved spherical lens (a meniscus) at the front of the tube. The secondary mirror in the MCT is a small spot aluminized directly onto the interior surface of the meniscus. Light enters the tube through the meniscus, refracts onto the primary mirror and is reflected into the eyepiece via the secondary mirror and a central hole in the primary. Although MCTs superficially resemble SCTs, MCTs tend to be far better performers for the Moon and planets. With their long focal lengths and excellent correction for spherical aberration, MCTs deliver excellent-resolution, high-contrast views of the lunar surface.

Telescopic Resolution

So much detail can be seen on the Moon from the Earth that some people are under the impression that features as small as a house can readily be discerned (I speak from experience). I would be ecstatically happy if my telescope allowed me to glimpse lunar features as big as the pyramids at Giza: even this degree of resolution is at the very limits of what can be achieved from the Earth, even through the largest professional telescopes. The bigger a telescope's objective lens or primary mirror, the finer the detail that will be seen on the Moon; this, however, is

Table 9.1. Aperture, resolving power and the limits of magnification

Aperture (mm)	Resolution (arcsec)	Smallest crater (km)	Suggested max magnification
30	3.8	7.2	60
40	2.9	5.5	80
50	2.3	4.4	100
60	1.9	3.6	120
80	1.4	2.7	160
100	1.2	2.3	200
150	0.8	1.5	300
200	0.6	1.1	400
250	0.5	1	500
300	0.4	0.8	600

1-arcsecond resolution equates to around 1.9 km on the Moon at its mean apparent angular diameter of 31 arcminutes.

ultimately restricted by the quality of the viewing conditions (see Table 9.1). On nights of really good viewing, the resolving power (R, in arcseconds) of a telescope of aperture diameter (D, in millimeters) can be calculated using the formula $R = 115/D$.

Seeing Conditions

From the surface of the Earth, we view space through a thick layer of atmosphere – 99% of the Earth's atmosphere lies in a layer just 31 km thick. It is the bottom 15 km of the atmosphere that causes the most problems. Clouds are the most obvious impediment to astronomical observation, but even a perfectly cloud-free sky can be absolutely useless for telescopic observation. The atmosphere is full of air cells of varying sizes (2–20 cm) and density, and light is refracted slightly differently as it passes through each cell. The worst viewing is produced when the air cells are mixing vigorously, making the light from a celestial object appear to jump around. The level of turbulence observed also varies with the height of the observed object above the horizon: a telescope that is pointed low will be looking through a far thicker slice of air than one pointed high. The observer's immediate environment also plays a significant role in how good the image is. A telescope brought out into the field needs some time to cool down. Chimneys, houses and factory roofs that give off heat produce columns of warm air that mix with the cold night air to warp the image.

Seeing

Seeing varies from 0.5-arcsecond resolution on an excellent night at a world-class observatory site to 10 arcseconds on the worst nights. On nights of poor visibility, it's hardly worth observing the Moon with anything but the lowest powers, since turbulence in the Earth's atmosphere will make the lunar surface appear to roll and shimmer, rendering any fine detail impossible to discern. For most of us, viewing rarely allows us to resolve lunar detail finer than 1 arcsecond, regardless of the size of the telescope used, and more often than not a 150mm telescope will show as much detail as a 300mm telescope, which has a light-gathering area 4 times as great. It is only on nights of really good visibility that the benefits of the resolving powers of large telescopes can be experienced. Unfortunately, such conditions occur all too infrequently for most amateur astronomers.

Eyepieces

A telescope can have the most perfectly figured lenses or mirrors, but it will not perform at its best if a poor-quality eyepiece is used. An eyepiece's magnification can be calculated by dividing the telescope's focal length by the focal length of the eyepiece. A 20mm eyepiece used on a telescope with a focal length of 1,500 mm will deliver a magnification of 75 (1,500/20 = 75). The same eyepiece will magnify ×40 on a telescope with a focal length of 800 mm (800/20 = 40).

For lunar observation, it is advisable to have at least three good-quality eyepieces that deliver low, medium and high-magnifications. A low-power eyepiece is excellent to introduce friends and family to the wonders of the Moon's surface. An eyepiece of 20mm focal length with a 50° apparent field will show an actual field twice the size of the Moon, at a magnification of ×50, using a telescope of 1000mm focal length. A high-power eyepiece should deliver a magnification that is double the telescope's aperture in millimeters – for example, ×200 on a 100mm refractor. High-powered lunar scrutiny can only be performed when viewing conditions allow. A medium-powered eyepiece that magnifies around ×50 (on telescopes smaller than 80mm) to ×100 (on larger telescopes) can be used to observe a satisfying amount of detail on the Moon, even under moderate viewing conditions.

Eyeglass wearers should consider the eye relief of any eyepiece they may want to purchase. Eye relief is the maximum distance from the eyepiece that the eye can comfortably be positioned at to see the full field of view. Eyepieces with long eye relief allow their wearers to observe in comfort without having to remove them to get the eye close to the lens. Some eyepiece designs have better eye relief than others.

Eyepieces are produced in three barrel diameters – 0.965-inch, 1.25-inch and 2-inch. The 0.965-inch eyepieces that come with many cheap small telescopes are usually made of plastic and are of a very unsophisticated design and poor optical quality. Good 0.965-inch eyepieces are hard to find these days, so it's better to upgrade to 1.25-inch ones. Most telescope focusers are built to accept 1.25-inch eyepieces, and some of them can accommodate 2-inch barrel ones as well. Two-inch barrel eyepieces can be hefty beasts with incredibly large lenses. They usually accommodate very-wide-angle, long-focal-length optical systems that are ideal for deep-sky observation.

Budget telescopes are usually supplied with Huygenian, Ramsden or Kellner eyepieces, which all have very restricted apparent fields of view – more a deep-sea diving experience than a spacewalk! Huygenian, Ramsden and Kellner eyepieces are unsuitable for lunar observation at high-magnification.

The Huygenian is a very old design, consisting of two plano-convex lenses; the convex sides both face the incoming light, and the focal plane lies between the two lenses. Huygenians are undercorrected (the rays from the outside zone of the lenses come to a shorter focus than those focused by the central parts), but the aberrations from each lens effectively cancel each other out. Huygenians deliver very small apparent fields of 30° (or even smaller), and are only suitable for use with telescopes with a focal length of f/10 or greater. Huygenians have poor eye relief.

Ramsdens are another very old design; like Huygenians, they consist of two plano-convex lenses, but both convex sides face each other (sometimes the lenses may be cemented together to provide better correction); the focal plane lies in front of the field lens (the lens that first intercepts the light). Ramsdens deliver flatter, slightly larger apparent fields of view than Huygenians, but they are prone to a greater degree of chromatic aberration, are poor performers on short-focal-length telescopes and offer poor eye relief.

Kellners are the least ancient of the three basic designs. Similar to the Ramsdens, their eye lens (the lens closest the eye) consists of an achromatic doublet. Kellners deliver better-contrast views than Huygenians or Kellners, with fields of about 40°, but annoying internal ghosting is invariably seen when viewing bright objects like the Moon. Like the Ramsden, the Kellner's focal plane is located just in front of the field lens, so any minute particles of dust that happen to land on the field lens will be seen as dark silhouettes against the Moon. Kellners with focal lengths longer than 15mm perform the best, while shorter focal lengths can produce a blurred effect around the edge of the field, along with chromatic aberration. Kellners have good eye relief.

A monocentric eyepiece consists of a meniscus lens cemented to either side of a biconvex lens. Despite having narrow apparent fields of view of around 30°, monocentrics deliver excellent, crisp, color-free, high-contrast images of the Moon, completely free from ghost images, and they can be used with low-focal-length telescopes.

Orthoscopic eyepieces comprise four elements, an achromatic doublet eye lens and a cemented triplet field lens. They produce a flat, aberration-free field, and deliver very good high-contrast views of the Moon. Their apparent field of view varies from around 30° to 50° and they have good eye relief.

Erfle eyepieces have multiple lenses (usually a set of two achromatic doublets and a single lens, or three achromatic doublets) that deliver a wide 70° field of view with good color correction. Erfles perform at their best when used with long-focal-length telescopes, and the best versions are of 25mm focal length and greater. However, the definition at the edge of the field of view tends to suffer, and the multiple lenses produce internal reflections and annoying ghost images when bright objects like the Moon are viewed.

Today's most popular eyepiece is the Plössl, with a four-element design that produces good color correction and an apparent field of view around 50°, which is flat and sharp up to the edge of the field. Plössls can be used with telescopes of very short focal length. Standard Plössls with long focal lengths have a good degree of eye relief. Lower-focal-length Plössls of standard design have poor eye relief, so they may be a little awkward to use for high-magnification views of the Moon, but versions are available with greater eye relief.

Modern demand for quality wide-field eyepieces has led to the development of designs such as the Meade UWAs, Celestron Axioms, Vixen Lanthanum Superwides and Tele-Vue Radians, Panoptics and Naglers. These all deliver excellently corrected images with wonderfully large apparent fields of view, and all have good eye relief. It's thrilling to view the lunar surface through an eyepiece whose apparent field of 60°, 70° or 80° or greater cannot all be taken in by the eye at once. The large

Fig. 9.2. Plossl eyepieces are the most commonly used of all today's ocular designs. They deliver a fairly uniformly focused field of view, with good contrast and little ghosting. Their reasonably wide apparent field of view (typically around 50° and wider) over a range of focal lengths makes them good all-round eyepieces for close-up scrutiny of lunar features at high magnification to wider views that encompass the entire Moon.

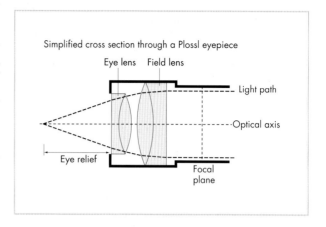

field of view can draw in the observer so completely that it is easy to imagine that there's no telescope between the observer and the Moon. With their 80°-plus apparent fields, Naglers are awesome eyepieces, and, like all quality super-wide-field eyepieces, they retail at an awesome price. A set of four Nagler eyepieces can cost more than a brand new 200mm SCT. Some of the older style long-focal-length Naglers are extremely big and heavy, and switching eyepieces requires rebalancing the telescope.

Zoom eyepieces remove the necessity of changing eyepieces of various focal lengths to vary the magnification. A number of reputable companies, including Tele-Vue, sell premium zoom eyepieces. Zoom eyepieces have been around for many years, but they have not yet achieved widespread popularity among serious amateur astronomers, perhaps because zooming is perceived to be a novelty associated with many budget binoculars and telescopes. Good quality zoom eyepieces are by no means joke items. They achieve a range of focal lengths by adjusting the distance between some of the lenses. A popular premium 8–24mm-focal-length zoom eyepiece has a narrow 40° apparent field when set at its longest focal length of 24mm, but as the focal length is reduced, the apparent field enlarges, up to 60° at 8mm. A good zoom eyepiece can replace a number of regular eyepieces and at a fraction of the cost. Features on the Moon can be zoomed in at leisure, but the telescope must be refocused each time the zoom is used.

Binocular Viewers

Binocular viewers split the beam of light from the telescope's objective into two components that are reflected into two identical eyepieces. Most binocular viewers require a long lightpath, and they will only work on an instrument whose focuser can be racked in enough for the prime focus to pass through the convoluted optical system of the binocular viewer. A binocular viewer may not be able to focus through a standard Newtonian telescope, and the best instruments are refractors and catadioptrics (SCTs and MCTs). Binocular viewers are designed to be used with two identical eyepieces (at the least, two eyepieces of the same focal length). Eyepieces of 25mm focal length and shorter are recommended, since vignetting around the edge of the field becomes apparent when much-longer-focal-length eyepieces are used. The use of two premium zoom eyepieces will prevent having to swap eyepieces to alter the magnification.

Viewing the Moon up close with two eyes, rather than one, has distinct advantages. It is more comfortable to use both eyes, and the view is more aesthetically pleasing. With two eyes, a 2-dimensional image takes on a near 3-dimensional appearance. Viewing with both eyes allows finer detail to be discerned.

Telescope Mounts

It is important that a telescope be attached to a sturdy mount, and that it is able to be moved without any difficulty in order to keep the Moon in the eyepiece as the Earth rotates. The simplest form of mount is a telescope inserted into a large ball that can freely and smoothly rotate in a cradle; several small reflectors are mounted in such a manner, and they are great fun to use.

Altazimuth Mounts

Altazimuth mounts enable the telescope to be moved up and down (in altitude) and from side to side (in azimuth). Small undriven tabletop altazimuth mounts are often provided with small refractors, but the quality of their construction can be poor. Most problems are caused by inadequate bearings on the altitude and azimuth axes; they may be too small and the right amount of friction may be difficult to achieve. Overly tight bearings result in too much force being used to overcome the friction, and smooth tracking cannot be achieved. Better altazimuth models are provided with slow-motion knobs that allow the telescope to be moved without having to push its tube around. If the mount itself is lightweight and shaky, it is liable to be buffeted by the slightest wind, rendering it unable to be used in the field. It may be better to attach the telescope to a good-quality camera tripod.

Dobsonians are altazimuth mounts that are used almost exclusively for Newtonian reflectors of short focal length. Since their invention several decades ago, they have become highly popular because they are simple to build and easy to use. Dobsonians consist of a ground box with an azimuth bearing and another box that holds the telescope tube. The altitude bearing is at the center of balance of the telescope tube, and it slips neatly into a recess in the ground box. Low friction materials like polythene, Teflon, formica and Ebony Star are used for the load-bearing surfaces, enabling the largest Dobsonians to be moved around at the touch of a fingertip. Lightweight structural materials such as MDF and plywood make Dobsonians strong but highly portable, and commercially produced Dobsonians range from 100mm to half-meter-aperture Newtonians.

It is not too difficult a task to keep the Moon in the field of view of a telescope mounted on an undriven altazimuth or Dobsonian mount up to magnifications of ×50. The higher the magnification, the faster the Moon appears to move across the field of view, and more frequent small adjustments need to be made to keep the Moon centered in the field. If the observer wants to make an observational sketch of lunar features, the limit of magnification for an undriven telescope is ×100 – anything higher and the instrument will need adjusting after each time the drawing is attended to, a tedious process that will double the length of time that a drawing should take to complete. At ×100, a feature that is centered in the field will take about 30 seconds to move to the edge of the field. High-magnification with undriven telescopes also demands a sturdy mount that does not shake unduly when pushed, and smooth bearings that respond to light touch and produce little backlash – qualities found in only the best altazimuth and Dobsonian mountings.

Equatorial Mounts

Serious lunar observation requires that a telescope to be mounted on a sturdy platform with one axis parallel to the rotational axis of the Earth and the other axis at right angles to it. In an undriven equatorial telescope, the Moon can be centered in the field of view and kept there with either an occasional touch on the tube or the turn of a slow-motion control knob that will alter the pointing of one axis – far easier than having to adjust the telescope on both axes of an altazimuth-mounted telescope to keep a celestial object within the field of view. A properly aligned, well-balanced driven equatorial allows the observer more time to enjoy the Moon without worrying about its quickly drifting out of the field of view. Equatorial clock drives run at a "sidereal" rate, enabling celestial objects centered in the field to remain there over extended periods of time, depending on how well the equatorial's polar axis is aligned, the accuracy of the drive rate and the apparent motion of the celestial object. Objects like close asteroids, comets and the Moon appear to move against the background stars during the course of an evening's observation. The Moon itself moves eastward on the celestial sphere by about its own diameter every hour. A telescope that drives at a sidereal rate, suitable for general astronomical observation, will not keep the Moon centered with high accuracy, and it will require occasional adjustment. Some clock drives are equipped with a button that alters the tracking from sidereal to lunar rate, and a high-end computerized telescope set up accurately will keep any object in its database centered in its field of view for many hours.

German equatorial mounts on aluminum tripods are most often used with medium to large refractors and reflectors. A telescope mounted on a German equatorial is able to be turned to any part of the sky, including the celestial pole. Schmidt-Cassegrain telescopes are commonly fixed to a heavy-duty fork-type equatorial mount. The telescope is slung between the arms of the fork, and the base is tilted to point to the celestial pole. When their visual backs have particularly large accessories attached, say, a CCD camera, these instruments are sometimes unable to view a small region around the celestial pole because the telescope cannot swing fully between the fork and the base of the mount. This is no problem for the lunar observer because the Moon never gets anywhere near the celestial pole!

Many amateurs choose to keep their mount and telescope in a shed and set it up whenever there is a clear night. Setting up requires some time, and is usually done in several stages. The mount's polar axis must be at least roughly aligned with the celestial north pole in order to track with any degree of accuracy. A tripod can be difficult to adjust on sloping ground, and the legs not only pose something of a navigational hazard when walking around the instrument in the dark, but the seated observer will invariably knock into the tripod from time to time, producing vibration of the image. To eliminate the time-consuming chore of setting up and polar aligning for each observing session, some amateurs construct a permanent pier set in concrete, upon which their German equatorial mount can be fixed and aligned, or to which their entire SCT and mount can be quickly and securely fixed.

Computerized Mounts

Computers are revolutionizing amateur astronomy in many ways, and one of the most visible is the increasing preponderance of computer-controlled telescopes. They come in all varieties – small refractors mounted on computer-driven altazimuth mounts to large SCTs on computerized fork mounts and German equatorials. Some standard, undriven equatorial mounts can be upgraded to accept either standard clock drives or computerized drives. When details of the observing location and the exact time are input, a computerized telescope can automatically slew to the position of any celestial object above the horizon at the touch of a few buttons on a keypad. Smaller computerized telescopes tend to have fairly insubstantial mounts that are incapable of supporting much more weight than the telescope itself while maintaining good pointing and tracking accuracy, while they are acceptable for visual lunar observation, they may not withstand the addition of a heavy accessory such as a digicam or binocular viewer. Larger computerized telescopes of the SCT varieties produced by Meade and Celestron, for example, are constructed well enough to accommodate hefty accessories. A computerized telescope can automatically slew to the position of the Moon and track it accurately at the touch of a button, and basic lunar information can be displayed on the keypad's viewscreen. Unsurprisingly, many lunar observers with conventional equatorial mounts may not find these small advantages great enough to persuade them to make the upgrade. It is often argued that computerized telescope mounts are leading to a dumbing-down of practical astronomy, since the convenience of being able to locate celestial objects eliminates the need for the amateur to learn his or her way around the skies and to star-hop to find the fainter deep-sky objects. Such a debate will doubtless continue long into the future.

Glossary

Albedo

A measure of an object's reflectivity. A pure white reflecting surface has an albedo of 1.0 (100%). A pitch-black, nonreflecting surface has an albedo of 0.0. The Moon is a fairly dark object with a combined albedo of 0.07 (reflecting 7% of the sunlight that falls upon it). The albedo range of the lunar maria is between 0.05 and 0.08. The brighter highlands have an albedo range from 0.09 to 0.15.

Anorthosite

Rocks rich in the mineral feldspar, making up much of the Moon's bright highland regions.

Aperture

The diameter of a telescope's objective lens or primary mirror.

Apogee

The point in the Moon's orbit where it is furthest from the Earth. At apogee, the Moon can reach a maximum distance of 406,700 km from the Earth.

Apollo

The manned lunar program of the United States. Between July 1969 and December 1972, six Apollo missions landed on the Moon, allowing a total of 12 astronauts to explore its surface.

Asteroid

A minor planet. A large solid body of rock in orbit around the Sun.

Banded crater

A crater that displays dusky linear tracts on its inner walls and/or floor.

Basalt

A dark, fine-grained volcanic rock, low in silicon, with a low viscosity. Basaltic material fills many of the Moon's major basins, especially on the near side.

Basin

A very large circular impact structure (usually comprising multiple concentric rings) that usually displays some degree of flooding with lava. The largest and most conspicuous lava-flooded basins on the Moon are found on the near side, and most are filled to their outer edges with mare basalts. The far-side basins are generally smaller and have minimal lava flooding, mainly at their centers.

Breccia

A composite rock made up of a variety of fragments formed as a result of high-energy impacts.

Caldera

A sizeable depression in the summit of a volcano caused by subsidence or explosion.

Capture hypothesis

A theory for the origin of the Moon suggesting that it originally formed as a planet in an independent orbit around the Sun but was later captured by the Earth's gravity.

Catena (plural: Catenae)

A chain of craters.

Central peak

An elevation found at the center of an impact crater, usually formed by elastic rebound of the lunar crust after impact.

Cleft

A small rille.

Co-accretion hypothesis

A theory that postulates that the Moon formed from a cloud of debris in orbit around the Earth. Also known as the "Sister Planet" theory.

Collision hypothesis

A theory of the formation of the Moon that appears to account for more idiosyncrasies of the Moon and its orbit than any other theory. It postulates that the Moon was formed from a cloud of material blasted out from the Earth after a glancing blow from a Mars-sized impactor. Popularly known as the "Big Whack" theory.

Colongitude

The selenographic longitude of the sunrise terminator. Tables of colongitude in an ephemeris are consulted in order to plan or research lunar observations.

Crater

A circular feature, often depressed beneath its surroundings, bounded by a circular (or near-circular) wall. Almost all the large craters visible on the Moon have been formed by aster-oidal impact, but a few smaller craters are endogenic, of volcanic origin.

Crescent Moon

The period between the New Moon and dichotomy when the earth-turned lunar hemi-sphere is less than half-illuminated.

Crypotomare

An ancient mare overlain and obscured by thick piles of ejecta from subsequent basin-forming impacts.

Dark halo crater (DHC)

A crater surrounded by a collar of dark material. In some cases this material is volcanic ash thrown out from a volcanic vent. Other DHCs are produced by impacts that excavate darker material from beneath the lunar surface.

Dark side

The hemisphere of the Moon not experiencing direct sunlight.

Dichotomy

Half-phase (first quarter or last quarter Moon).

Dome

A low, rounded elevation with shallow-angled sides. Most have been formed volcanically, but others are thought to have arisen as a result of subcrustal pressure.

Dorsum (plural: Dorsa)

Wrinkle ridge.

Earthshine

The faint, blue-tinted glow of the Moon's unilluminated hemisphere, visible to the naked eye when the Moon is a narrow crescent. It is caused by sunlight reflected onto the Moon by the Earth.

Eclipse

A phenomenon caused when the Moon passes directly in front of the Sun and casts a shadow onto the Earth (solar eclipse) or when the Moon moves through the Earth's shadow (lunar eclipse).

Ecliptic

The apparent path of the Sun against the celestial sphere during the year. The ecliptic is inclined by 23.5° to the celestial equator. The major planets follow paths close to the ecliptic, and the Moon's path inclines by some 5° to it.

Ejecta

A sheet of material thrown out from the site of a meteoroidal or asteroidal impact that lands on the surrounding terrain. Large impacts produce ejecta sheets composed of melted rock and larger solid fragments, in some cases producing bright ray systems. The brightness of the ejecta blanket gradually fades over time.

Elongation

The angular distance of the Moon or a planet from the Sun, viewed from the Earth, measured between 0° and 180° east or west of the Sun. For example, the first quarter Moon has an eastern elongation of 90°.

Endogenic

Having an internal origin. Lunar volcanoes and faults are endogenic.

Ephemeris

A table of numerical data or graphs that gives information about a celestial body in a date-ordered sequence, i.e., the rising and setting times of the Moon, the Sun's selenographic colongitude, etc.

Evection

A regular deviation of the Moon's orbital path around the Earth caused by the gravitational pull of the Sun.

Exogenic

Having an external origin. Most lunar craters are exogenic.

Far side

The hemisphere of the Moon that is constantly turned away from the Earth. The far side relates to all the features between 90° east and 90° west, but libration allows the terrestrial observer to glimpse some 59% of the Moon's surface over time.

Fault

A crack in the lunar crust caused by tension, compression or sideways movement.

First Quarter

Half phase between New Moon and Full, occurring one-quarter through the lunation.

Fission hypothesis

An old, now abandoned theory that attempted to explain the origin of the Moon as a chunk of material spun off from a rapidly revolving Earth.

Full Moon

When the lunar disc is completely illuminated by the Sun. Viewed from above, the Sun, Earth and Moon are in line.

Gibbous

The phase of the Moon between dichotomy and Full.

Graben

A valley created by crustal tension and bounded by two parallel faults. Examples can be observed around the edges of several maria.

Highlands

Heavily cratered regions of the Moon of generally higher elevation than the maria. They appear significantly brighter than the maria.

Impact crater

A pit in the Moon's crust formed by a solid projectile's striking the Moon at high speed, causing either a mechanically excavated crater (meteoroid impacts) or a large explosive excavation (asteroid impacts).

Lacus (Latin: Lake)

A small, smooth plain.

Lava

Molten rock extruded onto the surface by a volcano.

Limb

The very edge of the Moon.

Lithosphere

The Moon's solid crust.

Lunar

Pertaining to the Moon (from Luna, Roman goddess of the Moon).

Lunar eclipse

A period during which the Moon moves through the shadow of the Earth. Lunar eclipses can be penumbral, partial or total, and happen at Full Moon when the Sun, Earth and Moon are almost exactly in line.

Lunar geology

The study of the lunar rocks and the processes that sculpted the Moon's surface. Sometimes referred to as "selenology."

Lunation

The period taken by the Moon to complete one cycle of phases, from New Moon to New Moon, averaging 29d 12h 44m. This is the Moon's synodic month. Lunations are numbered in sequence from Lunation 1, which commenced on 16 January 1923. Lunation 1000 commenced on 25 October 2003.

Mare (Latin: sea. Plural: maria)

A large, dark, lunar plain. Maria fill many of the Moon's large multiringed basins and comprise a total of 17% of the Moon's entire surface area.

Massif

A large mountainous elevation, usually a group of mountains.

Mons (Latin: mountain. Plural: montes)

The generic term for a lunar mountain.

New Moon

The lunar phase during which all of the near side is unilluminated. Seen from above, the Moon lies directly between the Earth and Sun.

Occultation

The disappearance or reappearance of a star or planet behind the lunar limb.

Palus (Latin: marsh)

A small lunar plain.

Perigee

The point in the Moon's orbit where it is closest to Earth. At perigee, the Moon can be as close as 356,400 km from the Earth.

Promontorium (Latin: promontory)

A mountainous headland that projects into a lunar mare.

Ray

A bright feature of flat relief that radiates from many of the younger lunar impact craters. Part of the crater's ejecta system, bright-colored ray material also churns up the lunar surface to reveal lighter-colored material beneath.

Regolith

The upper layer of the Moon's surface, a mixture of compacted dust and rocky debris produced by eons of relentless meteoritic erosion.

Rift valley

A graben-type feature caused by crustal tension, faulting and horizontal slippage of the middle crustal block.

Rille

A narrow valley. Some rilles are linear or arcuate, caused by crustal tension or faulting. Others are sinuous, believed to have been caused by fast-moving lava flows.

Rima (Latin: fissure. Plural: rimae)

A rille.

Rupes (Latin: scarp)

A cliff produced by crustal tension, faulting and relative horizontal movement between the two crustal blocks.

Satellite

An object in orbit around a larger body. The Moon is Earth's only natural satellite.

Secondary cratering

Craters produced by the impact of large pieces of solid debris thrown out by a large impact. Secondary craters often occur in distinct chains, where piles of material have impacted simultaneously.

Seeing

A measure of the quality and steadiness of an image seen through the telescope eyepiece. Seeing is affected by atmospheric turbulence, especially by thermal effects.

Selene

Ancient Greek goddess of the Moon.

Selenology

The study of the history of the Moon's rocks and its surface processes. From Selene, ancient Greek goddess of the Moon.

Sinus (Latin: bay)

An indentation along the edge of a mare.

Synodic month

The period taken for the Moon to complete one cycle of phases from New Moon to New Moon.

Terminator

The line separating the illuminated and unilluminated hemispheres of the Moon. From New Moon to Full, we observe the morning terminator. From Full to New Moon, we see the evening terminator. The terminator creeps across the surface at a speed of about 0.5° lunar longitude per hour.

Transient Lunar Phenomena (TLP)

Rarely observed, short-lived, anomalous-colored glows, flashes or obscurations of local surface detail, whose causes are poorly understood. Also known as lunar transient phenomena (LTP).

Vallis (Latin: Valley)

A large trenchlike depression in the lunar crust.

Volcano

An elevated feature built up over time by the eruption of molten lava and ash. Lunar volcanoes are usually low, with shallow slopes and topped by tiny summit craters (vents). Volcanic activity on the Moon ceased more than 2 billion years ago.

Waning Moon

The period from Full Moon to New Moon, when the illumination of the Earth-turned lunar hemisphere decreases.

Wrinkle ridge

A linear or sinuous feature of low elevation traversing many of the marial plains. Some are lava flow fronts, others are features formed by compression as the mare surface contracted, while a few trace the buried outlines of features such as craters or inner basin rings. Also known as a dorsum.

Waxing Moon

The period from New Moon to Full Moon when the illumination of the Earth-turned lunar hemisphere increases.

Appendix: Resources

Societies

The Society for Popular Astronomy (SPA)
Website: http://www.popastro.com
Address: The Secretary, 36 Fairway, Keyworth, Nottingham, NG12 5DU, United Kingdom.
Email: membership@popastro.com
Founded in 1953, the SPA is the largest astronomical society in the UK. It is aimed at amateur astronomers of all levels. Publications include the quarterly magazine, Popular Astronomy and six News Circulars per year. The SPA hosts quarterly London meetings and free shows at the Tussaud's London Planetarium. Members can take advantage of a range of special member discounts and the SPA Book Scheme. Online joining facility is available. The SPA has a variety of Observing Sections, including an active Lunar Section (directed by the author since 1984) that has its own journal, *Luna*.

The British Astronomical Association (BAA)
Website: http://www.britastro.org
Address: The Assistant Secretary, The British Astronomical Association, Burlington House, Piccadilly, London, W1J 0DU, United Kingdom.
A UK-based astronomical association aimed at amateurs with an advanced level of knowledge and expertise. The BAA Lunar Section has an active Lunar Topographic Subsection directed by Colin Ebdon and its own excellent journal featuring members' observations, *The New Moon*.

The Royal Astronomical Society (RAS)
Website: http://www.ras.org.uk
Address: Royal Astronomical Society, Burlington House, Piccadilly, London, W1J 0BQ, United Kingdom.
Founded in 1820, the RAS is the UK's leading professional body for astronomy and astrophysics, geophysics, solar and solar-terrestrial physics and planetary sciences. Its bimonthly magazine, *Astronomy and Geophysics*, features occasional informative articles about the Moon. Fellowship of the RAS, however, is open to non-professionals also.

Association of Lunar and Planetary Observers (ALPO)
Website http://www.lpl.arizona.edu/alpo
ALPO Lunar Section website: http://www.lpl.arizona.edu/~rhill/alpo/lunar.html
This large association, based in the United States, has an active Lunar Topographic Section run by Bill Dembowsky. Its newsletter, The Lunar Observer, is an excellent publication that features members observational drawings and images. In addition, the ALPO Lunar Section has a number of coordinators specializing in areas such as dome research and transient phenomena.

American Lunar Society (ALS)
Website: http://otterdad.dynip.com/als
The ALS, based in the United States, is a group dedicated to the study of the Moon through observation and attention to current research. The ALS also aims to educate youngsters

through age-specific projects. The ALS has a great quarterly journal, *Selenology*, that features Moon-related articles and observations.

Unione Astrofili Italiani (UAI)
Website: http://www.uai.it/sez_lun/english.htm
Based in Italy, the UAI has an active lunar observing section, with a very informative English version of its website. UAI projects include lunar topography and transient phenomena (including lunar impacts).

Internet Resources

Apollo Image Atlas
Website: http://www.lpi.usra.edu/research/apollo/
A superb collection of Apollo images of the Moon from orbit. Fully searchable database.

Chuck Taylor's Lunar Observing Group on Yahoo
Website: http://groups.yahoo.com/group/lunar-observing/
An active web-based community for discussing all things lunar.

Chuck Wood's Moon
Website: http://cwm.lpod.org/
A compendium of lunar science and history, compiled by Charles Wood, one of the world's foremost authorities on the Moon. Wood is a columnist for *Sky and Telescope* magazine and author of The Modern Moon and the Lunar 100 list.

Consolidated Lunar Atlas
Website: http://www.lpi.usra.edu/research/cla/
A collection of pre-Apollo photographic images covering the Moon's near side at a variety of illuminations and librations, taken through large Earth-based telescopes.

Digital Lunar Orbiter Photographic Atlas of the Moon
Website: http://www.lpi.usra.edu/research/lunar_orbiter/
The entire collection of images returned from the US Lunar Orbiter probes of the 1960s, accompanied by useful labelled line maps of each image.

Geologic Lunar Research Group
Website: http://glrgroup.org/
An excellent Italian-based resource (with good English language text) for lunar observers interested in lunar geology and TLP research founded in 1997 by Raffaello Lena and Piergiovanni Salimbeni.

Lunar and Planetary Institute (LPI)
Website: http://www.lpi.usra.edu/
LPI, based in Houston, Texas, is a focus for academic participation in studies of the current state, evolution and formation of the Solar System. It has extensive collections of lunar and planetary data, an image-processing facility, an extensive library, education and public outreach programs, resources and products. The LPI also offers publishing services and facilities for workshops and conferences.

Lunarobservers.com
Website: www.lunarobservers.com
Peter Grego's website has plenty of information about the Moon and how to observe it and often features live webcasts of the Moon (including special events like lunar eclipses) and the brighter planets.

Lunar Photo of the Day
Website: http://www.lpod.org/
Daily lunar visual stimulation and education from Charles Wood – a new lunar image each day, accompanied by a detailed explanation.

NASA Lunar Exploration
Website: http://nssdc.gsfc.nasa.gov/planetary/lunar/apollo_25th.html
Links to descriptions of all the NASA lunar missions, robotic and manned.

Computer Programs

Lunar Map Pro
Publisher: RITI
Website: http://www.riti.com
An electronic Moon map capable of being zoomed in to high magnifications, with two modes – a detailed shaded map and an accurate vector graphic line rendition. Latest version includes 3-D terrain modeling.

Lunar Observer's ToolKit
Publisher: H Jamieson
Website: http://home.bresnan.net/~h.jamieson/
Designed to help lunar observers plan and record the Moon, this popular program has a comprehensive suite of handy tools.

LunarPhase Pro
Publisher: NovaSoft Ltd
Website: http://www.nightskyobserver.com
A complete Moon utility for Windows with a great range of capabilities. Features an interactive lunar atlas.

Virtual Moon Atlas
Publisher: Patrick Chevalley and Christian Legrand
Website: http://www.astrosurf.com/avl/UK_index.html
This freeware package is a must for the lunar observer who has a computer and wishes to plan or research observations. Packed with features, including a map of the entire Moon (switchable with a photographic map) and a geological lunar map.

Bibliography

A Portfolio of Lunar Drawings
By Harold Hill
Cambridge University Press, 1991. 240 pp.
Superb observations of a variety of lunar features by one of the world's most accomplished amateur lunar observers.

Epic Moon
By William P Sheehan and Thomas A Dobbins
Willmann-Bell, 2001. 363 pp.
A history of telescopic lunar exploration.

Exploring the Moon Through Binoculars and Small Telescopes
By Ernest H Cherrington Jr
Dover, 1984. 229 pp.
A well-written guide to the Moon and the features visible through the lunation.

Full Moon
Edited by Michael Light and Andrew Chaikin
Jonathan Cape, 1999. 243 pp.
Artist and photographer Michael Light drew on NASA's original photographic archives to assemble an archetypal lunar journey in images, from take-off to landing. Contains many gorgeous images of the lunar surface.

Patrick Moore on the Moon
By Patrick Moore
Cassell, 2001. 240 pp.
A well-written, entertaining and uncomplicated guide to all aspects of the Moon.

Mapping and Naming the Moon
By Ewen A Whittaker
Cambridge University Press, 1999. 242 pp.
A history of lunar cartography and nomenclature, written by the world's foremost authority on the subject.

Moon Observer's Guide
By Peter Grego
Philips (UK), Firefly (US), 2004. 192 pp.
Contains a complete guide to observing the Moon based on the daily progress of the terminator, illustrated with map sections.

Moonwatch
By Peter Grego
Philips (UK), Firefly (US), 2004.
Kit contains the book Moon Observer's Guide, Philips Moon Map and a poster of lunar phases.

Observing the Moon
By Peter T Wlasuk
Springer, 2000. 181 pp.
A solid and reliable guide on lunar observing techniques. Some great observations and images. CD-ROM included.

The Geology of Multi-Ring Impact Basins
By Paul D Spudis
Cambridge University Press, 1993. 263 pp.
A planetary geologist explains how the large lunar basins were formed.

The Modern Moon
By Charles Wood
Sky Publishing, 2004. 228 pp.
Wood's authoritative and clear text explains the variety of forces that have sculpted the Moon through the eons. The book draws on both traditional telescopic observations of the Moon and the modern explorations of the Apollo, Clementine and Lunar Prospector missions.

Atlas of the Lunar Terminator
By John E Westfall
Cambridge University Press, 2000. 292 pp.
A selection of amateur CCD images shows features along the terminator through the lunar month. Some of the images are over-processed and too contrasty to be of much use in seeing fine tonal detail, but it is a useful book to consult nonetheless.

Atlas of the Moon
By Antonin Rükl
Sky Publishing, 2004. 224 pp.
A wonderfully clear, detailed atlas of the Moon in 76 sections, showing most objects visible through a 100 mm aperture.

Philip's Moon Map
Drawn by John Murray
George Philip Ltd, 2004.
The near side of the Moon clearly drawn and labelled and indexed with more than 500 features. Text description by Peter Grego.

Photographic Atlas of the Moon
By S M Chong, Albert C H Lim and P S Ang.
Cambridge University Press, 2002. 145 pp.
Day-by-day photographic coverage of the whole Moon throughout the lunation.

The Hatfield Photographic Lunar Atlas
By Henry Hatfield.
Springer, 1998. 130 pp.
A collection of close-up lunar photographs taken between 1965 and 1967, showing areas under different angles of illumination. A very useful book to consult, even though modern CCD images show far more detail.

About the Author

Peter Grego has been a regular watcher of the night skies since 1976. He began studying the Moon in 1982. He observes from his garden in Rednal, United Kingdom, with a variety of instruments, ranging from a 127mm Maksutov to a 300mm Newtonian, but his favorite instrument is his 200mm SCT. Grego's primary interests are observing and imaging the Moon and bright planets, but he occasionally likes to 'go deep' during the dark of the Moon.

Grego has directed the Lunar Section of Britain's Society for Popular Astronomy since 1984. He edits three astronomy publications – *Luna* the Journal of the SPA Lunar Section, the *SPA News Circulars* and *Popular Astronomy* magazine. In addition, he writes and illustrates the monthly *MoonWatch* column in the UK's *Astronomy Now* magazine, and maintains his own website at www.luna robservers.com. Grego is the author of the books *Collision: Earth!* (Cassell, 1998), the *Moon Observer's Guide* (Philips/Firefly, 2004) and *Need to Know? Stargazing* (Collins, 2005). He is a Fellow of the Royal Astronomical Society.

Index of Lunar Features

Abenezra 205
Abenezra C 205
Abulfeda 205
Abulfeda E 205
Aestatis, Lacus 189
Aestuum, Sinus 149, 152, 154
Agarum, Promontorium 144
Agassiz, Promontorium 137
Agatharchides 174
Agrippa 131
Airy 203
Albategnius 202
Albategnius B 203
Albategnius KA 203
Albategnius Alpha 203
Aldrin 132
Aliacensis 205
Almanon 205
Al-Marrakushi 218
Alpes, Montes 109, 137
Alpes, Vallis 109, 118, 137
Alpetragius 114, 201
Alphonsus 114, 200
Alphonsus Alpha 200
Alphonsus, Rimae 201
Altai, Rupes 118, 207
Ammonius 200
Amoris, Sinus 144
Ampère, Mons 128
Amundsen 213
Anaxagoras 115, 134
Anaximander 157
Anaximander B 157
Anaximander D 157
'Ancient Newton' 159
Anguis, Mare 144
Apenninus, Montes 109, 125, 128
Apianus 205
Arago 132
Arago Alpha 132
Arago Beta 132
Archerusa, Promontorium 128
Archimedes 109, 129, 140
Archimedes, Montes 140–1
Archytas 135
Arduino, Dorsum 166
Argaeus, Mons 126
Argand, Dorsa 166
Argelander 203
Ariadaeus 131
Ariadaeus, Rima 131
Aristarchus 115
Aristarchus Plateau 162
Aristarchus, Rimae 166

Aristillus 109, 126, 140
Aristoteles 110, 136
Armstrong 132
Arzachel 114, 201
Arzachel A 202
Arzachel D 178
Arzachel E 202
Arzachel F 202
Arzachel, Rima 202
Asperitatis, Sinus 215
Atlas 147
Atwood 218
Australe, Mare 119, 224
Autolycus 109, 126, 140
Autumni, Lacus 189
Azara, Dorsum 126
Azophi 205

Baade 198
Baade, Vallis 198
Babbage 157
Babbage A 157
Baco 212
Baco B 212
Baillaud 134
Bailly 119, 181
Bailly A 182
Bailly B 182
Balboa 166
Balmer 218
Bancroft 150
Barkla 218
Barnard 219
Barocius 211
Barocius B 211
Barocius W 211
Barrow 134
Bartels 166
Beaumont 220
Beer 150
Behaim 218
Belkovich 147
Berosus 148
Bessel 126
Bessel D 126
Bianchini 161
Biela 222
Biela C 222
Bilharz 218
Billy 187
Birmingham 134
Birt 178
Birt A 178
Birt E 178

Birt F 178
Birt, Rima 178
Blanc, Mons 137
Blancanus 205
Blanchinus 205
Boguslawsky 213
Bohnenberger 220
Bohr 166
Bohr, Vallis 166
Bond, G 137
Bond, G, Rima 137
Bond, W 134
Bonitatis, Lacus 144
Bonpland 174
Boole 157
Borda 219
Boscovich 128
Boussingault 212
Boussingault A 213
Boussingault B 213
Boussingault C 213
Boussingault E 213
Boussingault K 213
Boussingault T 213
Bouvard, Vallis 198
Bradley, Mons 128
Bradley, Rima 129
Brayley 166
Brenner 222
Brianchon 157
Bucher, Dorsum 161
Buckland, Dorsum 126
Bulialdus 175
Bulialdus A 175
Burckhardt 144
Bunsen 171
Bürg 137
Bürg, Rimae 137
Burnet, Dorsa 165
Burnham 204
Byrd 134
Byrgius 191
Byrgius A 116, 191

Cabeus 214
Campanus 175
Capella 216
Capella, Vallis 216
Capuanus 180
Capuanus P 181
Cardanus 166
Cardanus, Rima 166
Carpatus, Montes 109
Carpenter 157

Casatus 185
Casatus C 185
Cassini 109
Cassini A 138
Catharina 115, 205
Catharina B 207
Catharina P 207
Catharina S 207
Cato, Dorsa 217
Caucasus, Montes 109, 110, 127
Cauchy 145
Cauchy, Omega 145
Cauchy, Rima 145
Cauchy, Rupes 145
Cauchy, Tau 145
Cavalerius 169
Cavendish 191
Cavendish E 191
Cayeux, Dorsum 217
Censorinus 216
Censorinus C 216
Cepheus 147
Chacornac 127
Challis 134
Cichus 180
Cichus B 180
Clairaut 212
Clairaut A 212
Clausius 197
Clavius 114, 184
Clavius C 184
Clavius D 184
Clavius J 184
Clavius JA 184
Clavius N 184
Clerke 127
Cleomedes 144
Cleomedes, Rima 144
Cognitum, Mare 111
Collins 132
Colombo 217
Colombo A 217
Compton 148
Concordiae, Sinus 145
Conon 129
Copernicus 114, 115
Cordillera, Montes 189
Cremona 157
Crisium, Mare 108, 143
Cruger 189
Curtius 213
Cushman, Dorsum 217
Cuvier 212
Cyrillus 115, 205
Cyrillus A 207
Cyrillus Alpha 207
Cyrillus Delta 207
Cyrillus Eta 207

Daguerre 216
Dalton 166
Damoiseau 188
Daniell, Rimae 137
Darney 173
Darwin 189
Darwin, Rimae 191

da Vinci 145
Davy 174
Davy A 174
Davy, Catena 174
Davy G 174
Davy Y 174
Dawes 126
de Gasparis 191
de Gasparis A 191
de Gasparis, Rimae 191
Delambre 204
de la Rue 135
Delaunay 205
Delisle 161
Delisle, Mons 161
Dembowski 131
Democritus 135
Demonax 213
Descartes 204
Descartes A 204
Descartes E 204–5
Descensus, Planitia 169
Deslandres 179, 180
Deville, Promontorium 137
Diophantus 180
Doloris Lacus 128
Doppelmayer 194
Doppelmayer K 191
Doppelmayer, Rimae 194
Draper 150
Drebbel B 197
Drebbel E 197
Drygalski 185

Eddington 166
Eichstadt 189
Einstein 166
Einstein A 166
Elger 181
Encke 168
Encke T 168
Endymion 114, 147
Epidemiarum, Palus 175, 179, 180
Epigenes 134
Eratosthenes 109, 150
Euclides 173
Euclides P 173
Euctemon 134
Eudoxus 110, 136
Euler 150
Ewing, Dorsa 187
Excellentiae, Lacus 197

Fabricius 222
Fabricius A 222
Faraday 210
Fauth 114
Fauth A 114
Fecunditatis, Mare 108
Felicitatis, Lacus 128
Fernelius 210
Feuilée 150
Firmicus 145
Flamsteed 187
Flamsteed G 187

Flamsteed P 187
Fontenelle 158
Fourier 194
Fracastorius 108, 220
Fracastorius D 220
Fracastorius E 220
Fracastorius H 220
Fracastorius M 220
Fracastorius Y 220
Fra Mauro 174
Fra Mauro E 174
Franklin 147
Fresnel, Promontorium 129
Frigoris, Mare 108
Furnerius 222
Furnerius A 216
Furnerius B 222
Furnerius, Rima 222

Galilei 169
Galilei, Rima 169
Galle 135
Galvani 171
Gambart 154
Gambart B 154
Gambart C 154
Gärtner 135
Gärtner, Rima 135
Gassendi 115, 192
Gassendi A 193
Gassendi, Rimae 193
Gast, Dorsum 126
Gaudibert 216
Gaudii, Lacus 128
Gauricus 180
Gauss 148
Gay-Lussac 152
Gay-Lussac, Rima 152
Geber 205
Geikie, Dorsa 217
Geminius 146
Gemma Frisius 211
Gemma Frisius A 211
Gemma Frisius B 211
Gemma Frisius C 211
Gerard 171
Gibbs 218
Gioja 134
Gilbert 218
Goclenius 217
Goclenius B 217
Goclenius, Rimae 217
Goclenius U 217
Goddard 145
Godin 131
Goldschmidt 134
Goodacre 211
Gould 177
Grabau, Dorsum 159
Greaves 143
Grimaldi 115, 187
Grimaldi, Rimae 188
Gruemberger 213
Gruithuisen 161
Gruithuisen Delta, Mons 108
Gruithuisen Gamma, Mons 108

Guericke 174
Guericke F 174
Gutenberg 216
Gutenberg C 216
Gutenberg E 216
Gutenberg, Rimae 216
Gylden 200

Hadley, Mons 128
Hadley Delta 128
Hadley, Rima 129
Haemus, Montes 128
Hagecius 212
Hagecius A 212
Hainzel 181
Hainzel A 181
Hainzel C 181
Hale 213
Hall 137
Halley 202
Hansteen 187
Hansteen, Mons 187
Hansteen, Rima 187
Harbinger, Montes 187
Harker, Dorsa 143
Harpalus 157
Hase 219
Hase D 219
Hase, Rima 222
Hausen 182
Hayn 147
Hecataeus 218
Hecataeus K 218
Hedin 169
Heim, Dorsum 160
Heinsius 183
Helicon 160
Hell 180
Helmholtz 213
Henry 191
Henry Frères 191
Heraclides, Promontorium 109
Heraclitus 212
Heraclitus D 212
Hercules 114, 147
Herigonius 187
Herigonius, Rima 187
Hermite 134
Herodotus 164
Herodotus Mons 165
Herodotus Omega 165
Herschel 200
Herschel, C 160
Herschel, J 157
Hesiodus 175
Hesiodus A 175
Hesiodus, Rima 177
Hevelius 169
Hevelius, Rimae 169
Hiemalis, Lacus 128
Higazy, Dorsa 150
Hind 204
Hippalus 195
Hippalus, Rimae 195
Hipparchus 202
Hipparchus C 204

Hipparchus L 204
Hipparchus X 202
Hohmann 189
Holden 218
Hommel 212
Hommel A 212
Hommel C 212
Hommel H 212
Horrocks 202
Hortensius 155
Hubble 145
Huggins 183
Humboldt 219
Humboldtianum, Mare 118
Humorum, Mare 108, 191
Huygens, Mons 128
Hyginus 130
Hyginus, Rima 130

Ibn Rushd 207
Il'in 189
Imbrium, Mare 107, 126
Inghirami 198
Inghirami, Vallis 198
Insularum, Mare 153
Iridum, Sinus 108
Isidorus 216
Isidorus A 216
Isidorus B 216

Jacobi 212
Janssen 222
Janssen, Rimae 222
Jenner 224
Julius Caesar 128
Jura, Montes 108

Kaiser 210
Kapteyn 218
Kästner 218
Kelvin, Promontorium 194
Kelvin, Rupes 194
Kepler 115
Kies 175
Kies Pi 175
Kiess 218
Klaproth 185
Klein 203
König 175
Kopf 189
Krafft 166
Krafft, Catena 166
Krustenstern 205
Kuiper 174
Kundt 174
Kunowsky 168

la Caille 205
Lade 204
Lagrange 198
Lalande 174
Lamarck 191
Lamb 224

Lambert 150
Lambert R 150
Lamé 218
Lamé G 218
Lamé P 218
Lamont 111, 132
Langrenus 218
Langrenus Alpha 218
Langrenus Beta 218
Langrenus DA 218
Lansberg 173
Lansberg C 173
Lansberg D 173
Langrenus V 218
la Pérouse 218
Laplace, Promontorium 160
Lawrence 145
Lavoisier 171
Lavoisier A 171
Lee 194
Lee M 194
le Gentil 182
Lehmann 198
Lehmann E 197
le Monnier 127
Lenitatis, Lacus 128
Letronne 187
Letronne B 187
Letronne W 187
Letronne X 187
le Verrier 160
Lexell 180
Licetus 212
Lichtenberg 171
Liebig 191
Liebig, Rupes 193
Lilius 212
Lilius A 212
Lilius C 212
Lindenau 211
Linné 126
Lister, Dorsa 126
Littrow 127
Littrow, Rimae 127
Loewy 195
Lohrmann 169
Lohse 218
Longomontanus 183
Longomontanus Z 184
Lorentz 171
Louville 161
Lubbock 217
Lubiniezky 174
Lyot 224

Maclear, Rimae 132
Macrobius 144
Mädler 216
Maestlin 168
Maestlin R 168
Maestlin, Rimae 168
Magelhaens 217
Magelhaens A 217
Maginus 184
Main 134
Mairan 161

Mairan, Rima   161
Mallet   222
Manilius   130
Manzinus   212
Marco Polo   130
Marginis, Mare   145
Marinus C   222
Marius   168
Marius Hills   168
Marius, Rima   169
Marth   181
Maskelyne A   216
Mason   137
Maunder   189
Maupertuis   161
Maupertuis, Rimae   161
Maurolycus   210
Mawson, Dorsa   217
Mayer, T   152
Medii, Sinus   131
Mee   181
Menelaus   126
Mercator   175
Mercator, Rupes   175
Mersenius   193
Mersenius C   194
Mersenius D   194
Mersenius, Rimae   194
Messala   146, 148
Messier   216–17
Messier A   217
Meton   134
Metius   222
Metius B   222
Milichius   155
Milichius Pi   155
Milichius, Rima   168
Miller   183
Mitchell   136
Montanari   183
Moretus   213
Moro, Mons   173
Mortis, Lacus   134
Moseley   166
Mösting   154
Moulton   224
Murchison   131
Mutus   212

Nasireddin   183
Nasmyth   198
Naonubu   218
Naumann   171
Nearch   212
Nearch A   213
Nectaris, Mare   108
Neison   134
Neper   145
Neumayer   213
Newton   214
Newton A   214
Nicol, Dorsum   126
Nicollet   177
Nielsen   171
Niggli, Dorsum   165
Nonius   210

North pole   134
Nubium, Mare   108

Odii, Lacus   128
Oenopides   157
Oken   224
Olbers A   169
Opelt   177
Oppel, Dorsum   143
Orientale, Mare   118, 189
Orontius   183
Owen, Dorsum   126

Palisa   174
Palitzsch   219
Palitzsch, Vallis   219
Pallas   131
Palmieri   191
Palmieri, Rimae   191
Parrot   203
Parry   174
Parry, Rimae   174
Pascal   157
Peary   134
Peirce   143
Penck, Mons   208
Perseverantiae, Lacus   145
Petavius   115, 218
Petavius A   219
Petavius B   216, 219
Petavius, Rimae   219
Philolaus   157
Phocylides   198
Piazzi   198
Picard   143
Piccolomini   207
Pico, Mons   109
Pico Beta, Mons   159
Pictet   183
Pictet E   183
Pitatus   175
Pitatus, Rimae   175
Pitiscus   212
Pitiscus A   212
Piton, Mons   109, 140
Plana   137
Plato   109, 115
Playfair   205
Playfair G   205
Plinius   128
Plinius, Rimae   128
Poisson   205
Pomortsev   217
Poncelet   158
Posidonius   127
Posidonius A   127
Posidonius B   127
Posidonius D   127
Posidonius J   127
Posidonius, Rimae   127
Prinz   166
Prinz, Rimae   166
Procellarum, Oceanus   107–8
Proclus   115, 143
Protagoras   135

Ptolemaeus   114, 200
Ptolemaeus B   200
Puiseux   194
Purbach   205
Purbach G   205
Purbach H   205
Purbach L   205
Purbach M   205
Purbach W   205
Putredinis, Palus   129
Pyrenaeus, Montes   216
Pythagoras   157
Pytheas   150

Rabbi Levi   211
Raman   165
Ramsden   181
Ramsden, Rimae   181
Recta, Rupes   118, 177
Recti, Montes   109
Regiomontanus   205
Regiomontanus A   205
Reimarus   222
Reiner   169
Reiner Gamma   169
Reinhold   154
Repsold   171
Repsold, Rimae   171
Rhaeticus   131
Rheita   222
Rheita, Vallis   222
Riccioli   188
Riccioli, Rimae   188
Riccius   212
Riphaeus, Montes   111
Ritter   132
Ritter, Rimae   132
Rocca   188
Rocca A   191
Röntgen   171
Rook, Montes   189
Roris, Sinus   156
Rosenberger   212
Rosenberger D   212
Rosse   220
Rost   181
Rozhdestvenskiy   134
Rubey, Dorsa   187
Rümker, Mons   111
Russell   166
Rutherford   184

Sabine   132
Sacrobosco   205
Santbech   220
Sasserides   183
Saunder   204
Saussure   183
Scheiner   185
Schiaparelli   165
Schickard   115, 197
Schickard A   198
Schickard B   198
Schickard C   198
Schiller   181

Schiller-Zucchius impact basin 180
Schlüter 188
'Schneckenberg' ('snail mountain') 130
Schomberger 213
Schrödinger 224
Schrödinger, Vallis 224
Schröter 154
Schröteri, Vallis 162
Scilla, Dorsum 171
Scoresby 134
Scott 213
Secchi 145
Secchi, Montes 145
Segner 181
Seleucus 166
Serenitatis, Mare 108, 125-6
Shaler 198
Sharp 161
Sharp, Rima 161
Sheepshanks 135
Sheepshanks, Rima 135
Short 214
Short A 214
Short B 214
Sikorsky 224
Silberschlag 131
Simpelius 213
Simpelius C 213
Simpelius D 213
Sirsalis 191
Sirsalis A 191
Sirsalis F 191
Sirsalis, Rima 191
Sirsalis Z 188
Smirnov, Dorsa 111, 126
Smythii, Mare 218
Snellius 219
Snellius, Vallis 219
Sömmering 154
Somni, Palus 145
Somniorum, Lacus 114, 127, 137
Sosigenes, Rimae 132
South 157
South pole 34, 214
Spei, Lacus 148
Spitzbergen, Montes 109, 141
Spörer 200
Spumans, Mare 217
Stadius 152
'Stag's Horn Mountains' 178

Steinheil 222
Stevinus 216
Stevinus A 216
Stiborius 212
Stöfler 210
Stöfler F 210
Stöfler K 210
Stöfler P 210
Strabo 135
Struve 166

Taenarium, Promontorium 177
Taruntius 145
Taurus, Montes 111, 127
Temporis, Lacus 146-7
Teneriffe, Montes 109
Tetyaev, Dorsa 143
Thales 115, 135
Thebit 178
Thebit A 178
Thebit L 178
Thebit P 178
Theon Junior 204
Theon Senior 204
Theophilus 115, 205
Theophilus B 206
Theophilus Alpha 206
Theophilus Phi 206
Theophilus Psi 206
Timaeus 134
Timocharis 150
Timoris, Lacus 181
Tisserand 144
Tolansky 174
Torricelli 216
Torricelli R 216
Toscanelli 165
Toscanelli, Rupes 165
Tralles 144
Tranquillitatis, Mare 108, 125-6
Tranquillitatis, Statio 132
Triesnecker 131
Triesnecker, Rimae 131
Tycho 115, 182
Tycho A 183
Tycho X 183

Ulugh Beigh 170-1
Undarum, Mare 142, 145, 218

'Valentine Dome' 127
Vaporum, Mare 108, 125, 130
Vasco da Gama 166
Vendelinus 218
Veris, Lacus 189
Very 126
Vieta 194
Vinogradov, Mons 150
Vitello 194
Vitruvius, Mons 126
Vlacq 212
Vogel 204
Vogel A 204
Von Braun 171
von Cotta, Dorsum 126
Voskresenskiy 166

Wallace 150
Walter 209
Walter A 210
Walter E 210
Walter K 210
Walter L 210
Walter W 180
Wargentin 198
Wargentin A 198
Watt 222
Webb 217
Webb C 217
Weigel 181
Weigel B 181
Werner 205
Whiston, Dorsa 171
Wichmann 187
Wichmann R 187
Widmanstätten 218
Wilhelm 183
Winthrop 187
Wolf 177
Wrottesley 219
Wurzelbauer 180

Yerkes 143
Young 222

Zagut 211
Zeno 148
Zupus 187
Zupus, Rimae 187

Subject Index

Achromatic objectives, 240
Afocal photography, 86–87
Age, phase and, 83, 51–52
Agrippa, 131
Albategnius, 203
Albedo, 50
Albedo features, 36
Alphonsus, 201
Altazimuth mounts, 247–248
Anomalistic month, 48
Anorthosite, 9, 17, 18
Antoniadi scale, 82
Apochromatic refractors, 240
Apogee, 43–44
Archimedes, 141
Arcuate rilles, 32
Aristarchus, 26, 163
Aristarchus plateau, 164
Aristarchus-type banded crater, 227–228
Aristillus, 26
Aristoteles, 135
Asteroidal impacts, 10
Asteroids, 66
Atlantic Ocean, 5
Atmospheric effects, 102–103
Atmospheric phenomena, 54
Averted vision, 232

Bailly, 182
Baily's Beads, 56
Banded craters, 227–230
Barringer crater, 64
Barycenter, 43
Basalt, 16
Basin flooding, 28–29
Basins
—impact, 27
—multiringed, 27–28
—ringed, 27
"Big Whack" theory, 8–9
Binocular viewers, 246–247
Binoculars, 104, 234–236
—image-stabilized, 235
—Porro prism, 236
—roof prism, 236
Birt-type banded crater, 229
Blind spots, 233
Blinking moon, 226–227
Breccia, 17, 19
Burg, 136

Callisto, 68
Caloris Planitia, 61–62

Camcorders, 93–94
Capture theory of Moon's formation, 6–8
Cassegrain reflectors, 241
Catadioptrics, 242
Catharina, 206, 207
Chicxulub crater, 64
Clavius, 184
Clementine topographic map, 21
Cleomedes, 144
Co-accretion theory of Moon's formation, 5–6
Collimation, 241
Collision theory of Moon's formation, 8–9
Color perception in moonlight, 102
Comets, 66–67
Computerized mounts, 249
Cone cells, 232
Conon-type banded crater, 228–229
Conventional photography, 85–86
Copernican Period, 12
Copernicus, 26, 114, 150, 152
Copied drawings, 80–81
Corona, lunar, 54, 103
Cosmic impact, 9
Crater depths, calculating, 230–231
Craters
—banded, 227–230
—impact, 18–28, 113–115
Cross-staff, lunar, 101
Crustal stresses, 38
Cryptomaria, 29
Cyrillus, 206, 207

Danjon scale, 119
Dark-halo craters, 33
Darwin, observational drawing of, 190
Deimos, 65–66
Delambre, 204
Delisle, 161
Deslandres, 180
Digital cameras, 92–93
Digital imaging, 92
Dione, 69
Domes, 29, 111, 113
Dorsa, 30, 111
Drawings, 77
—copied, 80–81
Dust transport, 40

Earth, 3, 60, 63–64
Earth-Mars centrifugal separation, 5

Earthshine, 51, 100
Eclipses, 54–57
—lunar, see Lunar eclipses
—solar, 55–56
Ecliptic, plane of, 44
Ejecta blanket, 21–23
Endymion, 148
Equatorial mounts, 248–249
Equipment, lunar observer's, 232–249
Eratosthenes, 150, 151
—observational drawing of, 151
Eratosthenian Period, 12
Erfle eyepieces, 245
Eudoxus, 135
Europa, 68
Evection, 46
Exposure times, 91–92
Eye, human, 232–233
Eye checkups, 233
Eyepiece projection, 91
Eyepieces, 240, 244–246

Fault planes, 32
Faults, 31–32
—finding, 116–118
Feature heights, calculating, 230–231
Feature names, 81
Film types, 88–89
Fission theory of Moon's formation, 4–5
Floaters, 233
Focal length, 238

Galilean telescopes, 239
Ganymede, 67–68
Gassendi, 192, 193
Gemma Frisius, 211
Ghost crater, 165
Graben rilles, 32–33
Gravity, tides and, 44–45
Gravity maps, 14–15
Greenwich Mean Time (GMT), 81
Grimaldi, 188

Halley's Comet, 66–67
Harvest Moon, 54
Hippalus, 194
Hippalus A-type banded crater, 230
Hipparchus, 202
Human eye, 232–233

Humboldt, 219
Huygenian eyepieces, 244–245

Iapetus, 69
Ice, lunar, 34–35
Image-stabilized binoculars, 235
Impact, cosmic, 9
Impact basins, 27
Impact craters, 18–28, 113–115
Ink stippling, 80
Intensity-estimate scale, 79
International Astronomical Union (IAU), 81
Io, 68
ISO rating, 88–89

Jupiter, satellites of, 67–68

Kellner eyepieces, 245
Kepler, 167
—rays from, 163
KREEP, 9–10, 11

Lagrangian points, 58
Lamont, 132
Lava tubes, 31
Libration, 34, 46–50, 83
—optical, 46–50
—physical, 46, 50
Libration features, 118–119
Lighting effects, 50–53
Line drawing, 78–79
Line of apsides, 44
Linné, 39
Lunar data, 85
Lunar eclipses, 56–57
—imaging, 121
—observing, 119
Lunar observer's equipment, 232–249
Lunar research, advanced, 225–231
Lunar rock, types of, 15–18
Lunar showcase, 104–121
Lunations, numbering, 85

Magnetic field, 13–14
Maksutov-Cassegrain telescope (MCT), 242
Maps of Moon
—binocular-visible features, 107
—craters and ray systems, 113
—general, 105
—mountain ranges and peaks, 110
—northeastern quadrant, 125–148
    area one, 125–132
    area two, 133–141
    area three, 142–145
    area four, 146–148
—northwestern quadrant, 149–171
    area five, 149–155
    area six, 156–161
    area seven, 162–169
    area eight, 170–171

Maps of Moon (continued)
—record-breaking features, 106
—rilles, 117
—southeastern quadrant, 199–224
    area thirteen, 199–208
    area fourteen, 209–214
    area fifteen, 215–220
    area sixteen, 221–224
—southwestern quadrant, 172–198
    area nine, 172–178
    area ten, 179–185
    area eleven, 186–195
    area twelve, 196–198
—survey of near side, 123
—wrinkle ridges and domes, 112
Mare Australe, 223
Mare Crisium, 143
Mare Humboldtianum, 148
Mare Humorum, 192
Mare Imbrium, 139–140
Mare Imbrium, tour of, 108–111
Mare Nubium, 175
Mare Orientale, 189
Maria, 105–108
Marius, 168
Mars, 64–65
—satellites of, 65–66
Mascons, 14
Megaregolith, 12, 13
Mercury, 7, 60–62
Messier pair, 39
Messier-type banded crater, 229
Meteorites, lunar, 20
Meteoroid impacts, 37
Meteoroids, 21
Mg-suite rocks, 17
Micrometeorite impact pit, 24
Micrometeoroids, 20
Milichius, 154
Miranda, 70
Monochromatic eyepieces, 245
Mons Delisle, 161
Mons Rümker, 171
Montes Alpes, 138
Montes Carpatus, 152
Montes Caucasus, 127
Montes Jura, 160
Montes Riphaeus, 173
Moon
—blinking, 226–227
—crust, 12
—development, 9–12
—drawing, 76–81
—features visible on, 98–99
—imaging, 85–97
—inside, 12–15
—mantle, 12
—maps of, see Maps of Moon
—mass, 42
—mountains of, see Mountains of Moon
—observational information, 81–85
—observing and recording, 75–97
—orbit of, 42–44
—origin, 3–40
—size, 41–42
—surface of, see Surface of moon

Moon (continued)
—tracking, 101–102
—viewing with unaided eye, 98–103
Moon dogs, 103
Moon illusion, 100–101
Moonlight, color perception in, 102
Moonquakes, 12
Moretus, 214
Mountains of moon, 33–34
—calculating heights of, 230–231
Multiringed basins, 27–28

Neap tides, 45
Nectarian Period, 10
Neptune, satellites of, 70–71
Newtonian reflectors, 241–242

Oberon, 70
Occultations, lunar, 57–58
Opera glasses, 104
Optical libration, 46–50
Orientale basin, 25
Orthoscopic eyepieces, 245

Pacific Ocean, 5
Parselene, 54, 103
Pencil sketches, 77–78
Penumbra, 54–55
Perigee, 43–44
Perturbations, secular and periodic, 45–46
Phases, 51–52
—age and, 83
—observability of, 52–53
Phobos, 65–66
Photography
—afocal, 86–87
—conventional, 85–86
—prime focus, 90–91
—SLR, 88
Physical libration, 46, 50
Pickering scale, 82
Plato, 158
Plinius, 128
Plössl eyepieces, 245
Prime focus photography, 90–91
Prism binoculars, 236
Proteus, 70–71
Ptolemaeus, 200, 201

Rainbows, lunar, 54, 102
Ramsden eyepieces, 245
Ray systems, 23, 27, 115–116
Reflectors, 241–242
Refractors, 238–241
Regolith, 15–16
Retardations, 53, 54
Retina, 232
Rhea, 69
Rilles
—arcuate, 32
—graben, 32–33
—sinuous, 31

Rima Hyginus, 130
Rima Sirsalis area, 190
Rimae Hippalus, 194
Ringed basins, 27
Rupes Altai, 208
Rupes Recta, 32, 176, 177

Saturn, satellites of, 69
Schickard, 197
Schmidt-Cassegrain telescope
    (SCT), 242
Schrödinger, 25
Secondary craters, 23
Seeing conditions, 243
Selenographic colongitude, 84
Selenographic coordinates, 83–84
Shadow contact timings, 120
Shadowplay, 53
Sinuous rilles, 31
Sinus Iridum, 160
SLR photography, 88
Soil creep, 39
Solar eclipses, 55–56
Solar System, 60, 71
South polar region, 213
Spring tides, 45
Stöffler region, 210
Summer solstice, 52
Surface of moon
—changes on, 35–40
—physical changes on, 38–40
—projection of entire, 28

Surface of moon (continued)
—shaping, 18–35
Synodic month, 52
Syzygy, 45

Tectonic activity, 19
Teleconverters, 89
Telephoto lenses, 89–90
Telescope mounts, 247–249
Telescopes, 236–246
Telescopic orientation, 122
Telescopic resolution, 242–244
Temperature change, 37–38
Terminator, 53
Terminology, lunar, 124
Terrestrial Dynamical Time (TDT),
    58
Tethys, 69
Theophilus, 206, 207
Thermal shock, 38
Thermoluminescence, 36
Tidal cycle, 45
Tidal forces, 5
Tides, gravity and, 44–45
Titan, 69
Titania, 69–70
TLP (transient lunar phenomena),
    35–36, 225–226
Tonal sketches, 77–78
Transient lunar phenomena (TLP),
    35–36, 225–226
Transparency, 83

Triton, 70
Tycho, 25, 26, 27, 183

Umbra, 54–55
Umbriel, 70
Universal Time (UT), 81
Uranus, 9
—satellites of, 69–70

Variation, 46
Venus, 62–63
Vieta, 26
Vignetting, 87
Visibility, 243–244
Vitamin E, 233
Volcanic activity, 11
Volcanoes, lunar, 29–30

Water, 16
Webcams, 94–97
Winter solstice, 52
Wolf Creek Crater, 64
Wood's Spot, 36
Wrinkle ridges, 30, 31, 111

Xenoliths, 30

Zoom eyepieces, 246